# Introduction to Risk and Failures

*Tools and Methodologies*

# Introduction to Risk and Failures

*Tools and Methodologies*

D. H. Stamatis

CRC Press
Taylor & Francis Group
Boca Raton   London   New York

CRC Press is an imprint of the
Taylor & Francis Group, an **informa** business

CRC Press
Taylor & Francis Group
6000 Broken Sound Parkway NW, Suite 300
Boca Raton, FL 33487-2742

First issued in paperback 2017

© 2014 by Taylor & Francis Group, LLC
CRC Press is an imprint of Taylor & Francis Group, an Informa business

No claim to original U.S. Government works

Version Date: 20140305

ISBN 13: 978-1-4822-3479-4 (hbk)
ISBN 13: 978-1-138-07191-9 (pbk)

**Visit the Taylor & Francis Web site at**
**http://www.taylorandfrancis.com**

**and the CRC Press Web site at**
**http://www.crcpress.com**

*To*

*Jeanna*

# Contents

List of Figures ................................................................................................... xiii
List of Tables .....................................................................................................xv
Acronyms ........................................................................................................ xvii
Preface............................................................................................................... xix
Acknowledgments ......................................................................................... xxiii
Author.............................................................................................................. xxv
Introduction .................................................................................................. xxvii

**1. Risk** ................................................................................................................ 1
  General Definition ..................................................................................... 1
  Other Definitions ...................................................................................... 2
    Economic Risks..................................................................................... 3
    Health Risks.......................................................................................... 3
    Health, Safety, and Environment (HSE) Risks................................. 3
    Information Technology (IT) and Information Security Risks ........... 4
    Insurance Risks .................................................................................... 4
    Business and Management Risks ....................................................... 4
    Human Services Risks.......................................................................... 5
    High Reliability Organizations (HROs) ........................................... 5
    Security Risks ....................................................................................... 6
    Societal Risks ....................................................................................... 6
    Human Factors Risks........................................................................... 7
  Risk Assessment and Analysis ................................................................ 8
  Quantitative Analysis............................................................................... 8
  Fear as Intuitive Risk Assessment........................................................... 9
  Audit Risk ................................................................................................ 10
  Other Considerations ............................................................................. 10
  Risk versus Uncertainty......................................................................... 10
  Risk Attitude, Appetite, and Tolerance................................................ 11
  Risk as Vector Quantity ......................................................................... 12
  Disaster Prevention and Mitigation ..................................................... 12
  Scenario Analysis.................................................................................... 13
  Notes......................................................................................................... 15
  References ................................................................................................ 20
  Selected Bibliography............................................................................. 22

**2. Approaches to Risk** ................................................................................... 25
  Zero Mind-Set ......................................................................................... 25
  ALARP...................................................................................................... 27
    U.K. Statistics: Work Accidents Involving Young People
    between 1996 and 2001 .................................................................... 30

U.S. Statistics (2011) ............................................................................30
How to Tell if a Risk Is ALARP.........................................................31
Risk Leverage........................................................................................36
Failures, Accidents, and Hazards .....................................................37
ALARP Fallacies....................................................................................37
Example........................................................................................................39
Differentiating Risks ...........................................................................40
Major Risks........................................................................................40
Serious Risks.....................................................................................41
Minor Risks .......................................................................................41
Risk Priorities...................................................................................41
Reference .....................................................................................................43
Selected Bibliography ...............................................................................43

3. **Types of Risk Methodologies** .................................................................45
Qualitative Methodologies .......................................................................45
Preliminary Risk Analysis..................................................................45
Hazard and Operability (HAZOP) Studies......................................46
Failure Mode and Effects Analysis (FMEA) and Failure Mode
and Criticality Effects Analysis (FMCEA).......................................46
Advantage .........................................................................................50
Disadvantages...................................................................................50
General Comments ...........................................................................50
Tree-Based Techniques..............................................................................52
Fault Tree Analysis (FTA) ..................................................................52
Event Tree Analysis (ETA) .................................................................52
Cause–Consequence Analysis ...........................................................53
Management Oversight Risk Tree (MORT) Analysis .....................53
Safety Management Organization Review Technique (SMORT) ......53
General Comments ...............................................................................53
Methodologies for Analysis of Dynamic Systems ................................54
GO Method ...........................................................................................54
Digraph or Fault Graph .....................................................................54
Markov Analysis (MA).........................................................................55
Dynamic Event Logic Analytical Methodology (DYLAM) ...............56
Dynamic Event Tree Analysis Method (DETAM)...............................56
General Comments ...............................................................................57
Traditional Methodologies .......................................................................57
What-If Method ....................................................................................58
Checklist................................................................................................61
What-If and Checklist Combination .................................................65
Indexing.................................................................................................65
Interface Hazards Analysis ................................................................65
References .....................................................................................................67

**4. Preliminary Hazard Analysis (PHA)** ....................................................... 69
   Example PHA: Home Electric Pressure Cooker ............................. 76
      Severity and Probability.......................................................... 80
      PHA Limitations ....................................................................... 81
      Preventive and Corrective Measures ..................................... 81
   References .............................................................................................. 82
   Selected Bibliography ....................................................................... 82

**5. HAZOP Analysis** ............................................................................................ 83
   Overview................................................................................................ 83
   Definitions............................................................................................. 84
   Process ................................................................................................... 86
      Minimum Requirements ......................................................... 86
      Defining Risk............................................................................ 86
      Trigger Events........................................................................... 87
      Use of Analysis........................................................................ 88
   HAZOP Process ................................................................................... 90
      Definition................................................................................... 90
      Preparation................................................................................ 91
      Examination............................................................................... 92
      Documentation and Follow-Up ............................................. 93
   Detailed Analysis................................................................................ 95
      Sequence of Examination........................................................ 96
      Deviations from Design Intent................................................ 97
      Details of Study Procedure..................................................... 98
   Effectiveness Factors.......................................................................... 99
   Team....................................................................................................... 99
      Team Leader (Chairperson)................................................... 100
      Engineers................................................................................. 100
   Description of Process........................................................................ 101
      Relevant Guidewords ........................................................... 102
      Point of Reference Concept................................................... 102
      Screening for Causes of Deviations..................................... 104
      Consequences and Safeguards ............................................ 105
      Deriving Recommendations (Closure)................................ 106
      Conditions Conducive to Brainstorming............................ 106
      Meeting Records .................................................................... 106
      Meeting Questions................................................................. 107
      Follow-Up................................................................................ 108
      Computer HAZOP (CHAZOP) ............................................ 108
         Advantages and Disadvantages.................................... 110
      Human Factors HAZOP......................................................... 110
   Report ................................................................................................... 110
      Study Title Page..................................................................... 110
      Table of Contents................................................................... 111

Glossary and Abbreviations ............................................................. 111
Aim ..................................................................................................... 111
Guidewords ...................................................................................... 111
Summary of Main Findings and Recommendations ...................... 111
Scope of Report ................................................................................ 111
Description of Facility ..................................................................... 111
Team Members ................................................................................. 113
Methodology .................................................................................... 113
Overview .......................................................................................... 113
Analysis of Main Findings .............................................................. 114
Findings ............................................................................................ 114
Review ................................................................................................... 114
Input Documents ............................................................................. 114
Review Information Pack ................................................................. 115
Review Team Composition .............................................................. 115
Full-Time or Core Team .............................................................. 116
Part-Time Team (Contractors or Consultants Engaged as
Needed) ........................................................................................ 116
Preparation ....................................................................................... 116
Methodology .................................................................................... 117
Recommendations ............................................................................ 118
Success Factors ..................................................................................... 119
Before Study ..................................................................................... 119
Throughout Study ........................................................................... 120
After Study ....................................................................................... 120
Revisions .......................................................................................... 121
References .............................................................................................. 122
Selected Bibliography .......................................................................... 122

6. Fault Tree Analysis (FTA) ................................................................. 125
Overview ............................................................................................... 125
Benefits ............................................................................................. 128
General Construction Rules ................................................................. 128
References .............................................................................................. 132
Selected Bibliography .......................................................................... 132

7. Other Risk and HAZOP Analysis Methodologies ........................... 133
Process Flowchart ................................................................................. 133
Functional Flow or Block Diagram ..................................................... 134
Advantages and Disadvantages .................................................... 136
Sketches, Layouts, and Schematics .................................................... 136
Failure Mode Analysis (FMA) ............................................................ 137
Control Plan .......................................................................................... 137
Process Potential Study (PPS) ............................................................. 138

Need and Feasibility Analysis ................................................................ 138
Task Analysis ............................................................................................ 138
    Advantages and Disadvantages .................................................. 140
Human Reliability Analysis .................................................................. 140
    Advantages and Disadvantages .................................................. 141
Failure Mode and Critical Analysis ..................................................... 141
Hazard Identification (HAZID) ............................................................ 142
    Phase 1: Planning ......................................................................... 142
    Phase 2: Identifying Hazards ...................................................... 143
    Phase 3: Evaluating Hazards ....................................................... 144
    Phase 4: Assessing Risks ............................................................. 144
    Phase 5: Managing Risks ............................................................. 145
    Phase 6: Monitoring Risks ........................................................... 146
Crisis Intervention in Offshore Production (CRIOP) ......................... 146
Hazard Analysis and Critical Control Points (HACCP) .................... 146
Near-Miss Reporting .............................................................................. 148
Incident and Accident Investigation and Reporting ......................... 149
Semi-Quantitative Risk Assessment (SQRA) ...................................... 150
Audits ....................................................................................................... 150
Event Tree Analysis (ETA) ..................................................................... 150
    Characteristics ............................................................................... 154
    Process ............................................................................................ 155
    Advantages and Disadvantages .................................................. 156
    Example .......................................................................................... 157
References ................................................................................................. 159
Selected Bibliography ............................................................................ 160

**8. Teams and Team Mechanics** ................................................................ 161
Team Members, Qualifications, and Activities ................................... 161
Benefits of Using Teams ........................................................................ 163
HAZOP Team ........................................................................................... 164
    Technicians .................................................................................... 164
    Mid-Level Managers .................................................................... 164
    Senior Managers ........................................................................... 165
Consensus ................................................................................................ 165
Team Process Check ............................................................................... 167
    Difficult Team Members ............................................................. 167
Problem Solving ...................................................................................... 169
Meeting Planning .................................................................................... 170
In-Process Meeting Management .......................................................... 172
Common Meeting Pitfalls ...................................................................... 173
Utilizing Meeting Management Guidelines ........................................ 174
References ................................................................................................. 175

9.  **OSHA Job Hazard Analysis** ................................................................... 177
    Reference .................................................................................................. 186
    Selected Bibliography ............................................................................ 186

10. **Hazard Communication Based on Standard CFR 910.1200**............... 187
    Hazard Communication Program and Hazardous Materials
    Control Committee .................................................................................. 188
        Members ............................................................................................... 188
        Responsibilities................................................................................... 189
        Employee Training.............................................................................. 189
        Employee Access ................................................................................ 190
        Information Sources ........................................................................... 191
            Labels .............................................................................................. 191
            Safe Use Instructions .................................................................. 191
            Chemical Materials Lists............................................................. 193
            Material Safety Data Sheets........................................................ 193
    References .................................................................................................. 200

**Appendix A: Checklists**.................................................................................. 201

**Appendix B: HAZOP Analysis Example** ...................................................... 215

**Index**................................................................................................................ 229

# List of Figures

**Figure P.1** Typical view of risk. .......................................................................... xx

**Figure P.2** Relationship between risk and uncertainty. .............................. xx

**Figure I.1** Preliminary HAZOP. .................................................................. xxix

**Figure I.2** Selection of HAZOP process. ........................................................ xxx

**Figure I.3** Node selection. ................................................................................. xxxi

**Figure 3.1** Typical flow for generating FMEA. ............................................ 48

**Figure 3.2** Interconnectivity. ............................................................................ 66

**Figure 4.1** PHA overview. ................................................................................. 70

**Figure 5.1** P&ID of feed section of process. ................................................... 89

**Figure 5.2** Revised P&ID of feed section of process. .................................... 89

**Figure 5.3** HAZOP procedure flow .................................................................. 95

**Figure 5.4** Operation with deviations. ............................................................ 96

**Figure 5.5** Cooling water facility. .................................................................... 97

**Figure 6.1** Typical partial engine FTA diagram. ......................................... 126

**Figure 6.2** Relationship of FTA and FMEA. ................................................ 126

**Figure 6.3** FTA depiction of parallel system. .............................................. 130

**Figure 6.4** Typical block diagram. ................................................................. 131

**Figure 7.1** Simple flowchart. ........................................................................... 134

**Figure 7.2** Logic depiction used in functional diagrams. .......................... 135

**Figure 7.3** Block diagram with designated boundary line. ....................... 136

**Figure 7.4** Overview of ETA. ........................................................................... 152

**Figure 7.5** Generic ETA showing primary and secondary trees. ............. 152

**Figure 7.6** Generic ETA associated with FTA and propabilities. .............. 154

**Figure 7.7** Typical ETA showing individual events of success and failure. ..................................................................................................................... 155

**Figure 7.8** ETA and FTA relationship. .......................................................... 157

**Figure 7.9** Anti-flooding system. .................................................................... 158

**Figure 7.10** Reliability diagram of flooding system.............................. 158

**Figure 7.11** ETA reliability diagram with associated probabilities. ......... 159

**Figure 8.1** Team overview............................................................. 162

**Figure 8.2** Team performance factors. ........................................... 162

**Figure B.1** Drawing DOP 001, Rev. 1............................................. 218

**Figure B.2** Drawing DOP 001, Rev. 2. ........................................... 226

# List of Tables

**Table 1.1** ISO/IEC 27001 Clauses Related to Risk ............................................. 6

**Table 2.1** Typical Cash Valuation for Cost–Benefit Analysis ..................... 34

**Table 2.2** Screening Measures ..................................................................... 39

**Table 2.3** Risk Assessment Matrix: Hazard Probability ........................... 42

**Table 2.4** Categories of Risk Assessment Matrix ..................................... 43

**Table 3.1** Initial FMEA Documentation .................................................. 47

**Table 3.2** Typical HAZOP/FMEA Worksheet ........................................ 49

**Table 3.3** FMEA and FMECA Worksheet ............................................... 51

**Table 3.4** Topics for Generating Checklist Questions .......................... 62

**Table 3.5** Sample Chemical Storage Checklist ...................................... 63

**Table 4.1** Typical Severity and Probability Classifications ................. 71

**Table 4.2** Typical Source for PHA Checklist ........................................ 73

**Table 4.3** PHA Worksheet ...................................................................... 77

**Table 4.4** Preliminary Hazard Matrix .................................................. 78

**Table 4.5** Typical PHA Brainstorming Record ..................................... 78

**Table 4.6** Typical PHA Report ............................................................... 79

**Table 5.1** A Simple Evaluation Method to Risk .................................. 87

**Table 5.2** HAZOP Steps ....................................................................... 91

**Table 5.3** HAZOP Recording Form ..................................................... 93

**Table 5.4** Typical Guidewords ........................................................... 103

**Table 5.5** HAZOP Meeting Record ................................................... 107

**Table 5.6** Guidewords and Parameters ............................................. 112

**Table 5.7** Revisions and Recommendations ..................................... 121

**Table 6.1** Typical FTA Symbols ......................................................... 127

**Table 6.2** FTA Logic Symbols ............................................................ 127

**Table 7.1** Typical Hazards Outside Envelope of Process Equipment ....... 143

**Table 7.2**  Worksheet for HAZID with SQRA ................................................ 151

**Table 7.3**  Flow of ETA .................................................................................... 153

**Table 7.4**  Flow of ETA in Application Format .............................................. 153

**Table 9.1**  Specific Hazards by Categories ...................................................... 178

**Table 9.2**  Hazard Categories and Controls ................................................... 180

**Table 9.3**  Hazard Analysis Form .................................................................... 181

**Table 9.4**  Hazard Analysis of Grinding Castings ......................................... 183

**Table 10.1**  Chemical Container Label ............................................................. 192

**Table 10.2**  Typical SUI Form .......................................................................... 194

**Table 10.3**  Generic Material Safety Data Sheet ............................................. 197

**Table A.1**  Safety Plan Checklist ...................................................................... 203

**Table A.2**  Facility Location Checklist ............................................................. 205

**Table B.1**  HAZOP Log Report ......................................................................... 220

# Acronyms

| Acronym | Meaning |
|---|---|
| ACOP | Approved Code of Practice |
| ACTS | Advisory Committee on Toxic Substances |
| ALARA | As Low as Reasonably Achievable |
| ALARP | As Low as Reasonably Practicable |
| CBA | Cost Benefit Analysis |
| CD | Consultative Document |
| CEN | Comité Européen de Normalisation |
| CENELEC | Comité Européen de Normalisation Electrotechnique |
| CLAW | Control of Lead at Work Regulations |
| COSHH | Control of Substances Hazardous to Health Regulations |
| CPF | Cost of Preventing a Fatality |
| CSF | Critical Safety Function |
| EC | European Communities |
| E/E/PE | Electrical, Electronic or Programmable Electronic |
| EU | European Union |
| FMRI | Final Mishap Risk Index |
| HSC | Health and Safety Commission |
| HSE | Health and Safety Executive |
| the HSW | The Health and Safety at Work, etc. Act |
| ICRP | International Commission on Radiological Protection |
| IEC | International Electrotechnical Commission |
| IMRI | Initial Mishap Risk Index |
| ISO | International Organization for Standardization |
| MEL | Maximum Exposure Limit |
| MHSWR | Management of Health and Safety at Work Regulations |
| NOAEL | No Observed Adverse Effect Level |
| OEL | Occupational Exposure Limit |
| OES | Occupational Exposure Standard |
| PHA | Process Hazard Analysis |
| PHL | Preliminary Hazard List |
| P&ID | Process and Instrumentation Diagrams |
|  | *Note*: In the chemical industry sometimes this acronym means Pipe and Instrumentation Diagrams) |
| PPE | Personal Protective Equipment |
| QRA | Quantitative Risk Assessment |
| RBMK | Reactor Bolshoi Mozjnoct Kanali |
| SFAIRP | So Far As Is Reasonably Practicable |
| SSR | System Safety Requirement |
| TLM | Top Level Mishap |
| TOR | Tolerability of Risk |
| VPF | Value for Preventing a Fatality |
| WATCH | Working Group on the Assessment of Toxic Chemicals |

# Preface

It has been said many times by many individuals that risk is everywhere. We can never avoid it. It is present in whatever we do. Obviously, we must try to understand the risks we face and minimize them if possible. This book is in fact an extension of my first edition published in 1995 and a second in 2003 on failure mode and effects analysis (FMEA) in which I discussed the benefits of prevention based on an up-front analysis of failures.

As time passed, I noticed that, whereas FMEA is a powerful tool to forecast failures of designs and processes, a missing link involving safety issues, catastrophic events, and their consequences had to be covered. The second edition briefly mentioned HAZOP analysis but did not expand on the methodology. In this book, I focus on risk and HAZOP as they relate to major catastrophic events and safety issues. Specifically, I address processes and implementation and explain the fundamentals of using risk methodology in any organization to evaluate major safety and/or catastrophic problems. A classical and typical view of risk is shown in Figure P.1.

The significance of Figure P.1 is that the risk is emphasized and indeed becomes more serious as both individual and societal risks become evident. In fact, the hidden and untold significance is that implicitly the figure also represents a level of uncertainty as shown in Figure P.2. Both risk and uncertainty in the final analysis may be viewed and analyzed from the following five perspectives (Callaghan and Walker 2001). In some cases, one factor may be predominant, but combinations of factors often must be identified and evaluated. The five perspectives are as follows:

**Individual concerns**—how individuals see the risk from a particular hazard affecting them, their families, and the things they value. While they may be prepared to engage voluntarily in activities that often involve high risks, as a rule they are far less tolerant of risks imposed on them and over which they have little control unless they see the risks as negligible. Moreover, while they may be willing to live with a risk that they do not regard as negligible that secures them or society certain benefits, they would want the risk levels low and clearly controlled.

**Societal concerns**—the risks or threats from hazards that impact society and, if realized, may produce adverse repercussions for the institutions responsible for putting in place the provisions and arrangements for protecting people through legislation. These concerns are often associated with hazards that give rise to risks that, if materialized, could provoke a socio-political response, for example, events causing widespread or large-scale consequences or multiple fatalities. Typical examples relate to nuclear power generation, transportation accidents, or the genetic modification of organisms. Societal concerns arising from multiple fatalities in a single

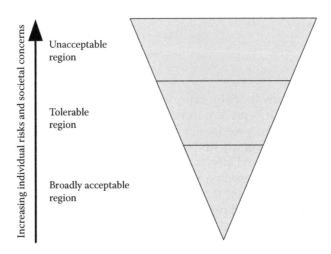

**FIGURE P.1**
Typical view of risk. (*Source:* www.HSE.gov.uk and public sector information published by the U.K. Health and Safety Executive and licensed under Open Government Licence v1. 0)

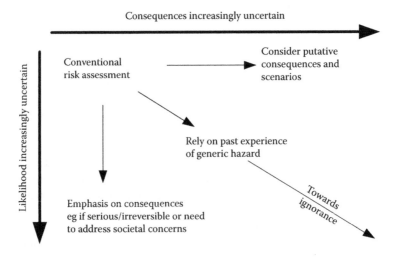

**FIGURE P.2**
Relationship between risk and uncertainty. (*Source:* www.HSE.gov.uk and public sector information published by the U.K. Health and Safety Executive and licensed under the Open Government Licence v1. 0)

event are known as *societal risks*. Societal risk is therefore a subset of societal concern.

**Complexity in government regulations**—regulations that affect and effect intra- and inter-state commerce and international commerce as well. Throughout the long history of legislation introduced to eliminate or minimize risks, the first areas to be regulated have always been the most obvious, often requiring little scientific insight for identifying problems and possible solutions. For example, it was not difficult to realize that controlling airborne dust would reduce the risk of silicosis in miners and that making it mandatory to guard moving parts of machinery would prevent workers from being killed or maimed. In short, dramatic progress toward tackling such problems could be (and was) made without unduly taxing existing scientific knowledge or the state of available technology. However, as the most obvious risks have been tackled, new and less visible hazards have emerged and gained prominence. Typical examples include hazards arising from biotechnology and processes that emit gases that contribute to global warming.

**Patterns of employment** defined by changing demographics present some challenges. The regulatory environment must cope with the increasing trend of industries to outsource work (and the attendant risks), resulting in changes in patterns of employment and in the fragmentation of large companies into autonomous organizations working closely together. Dramatic increases in self-employment and home working have been noted; small and medium size firms are now major forces in creating jobs. Moreover, many monolithic organizations have split into separate companies, for example, railways now operate as separate companies responsible for operating the tracks, rolling stock, and networks.

**Polarization of approaches between large and small firms** as a result of the patterns of employment. Some of these changes have blurred legal responsibilities for occupational health and safety, traditionally placed on those who created the risks and were best situated to control them. In certain industries, it has become difficult to determine who may be in that position. While case law clarified some situations, the fact remains that in many sectors it is very difficult to coordinate the adoption of measures to control risks. Many more players are involved, and some have little access to expertise.

Chapter 1 of this book serves as an introduction to risk and provides several definitions relevant to a number of industries. A distinction is also made between risk and uncertainty. Chapter 2 discusses approaches to risk and the zero mind-set philosophy. In conjunction with the concept of zero mind-set, the ALARP principle for determining what risk is and what its effects is also discussed. This chapter also addresses the major, serious, and minor categories of risks. Chapter 3 covers 18 risk methodologies dealing with analysis, failures, safety, and hazards.

Chapter 4 is about preliminary hazard analysis and explains how to evaluate a hazard in the early stages of a design. Chapter 5 covers hazard and operability (HAZOP) studies. It begins with an overview of HAZOP and provides key definitions. It also provides a detailed discussion of the study process, its effectiveness, and the team required to perform the study. It concludes with a full description of the process and report preparation. Chapter 6 focuses on fault tree analysis (FTA) and discusses the general rules of construction and the need for a top-to-bottom approach for defining failures and how they relate to HAZOP. Chapter 7 provides 14 additional risk analysis methodologies for handling HAZOP.

Chapter 8 is titled "Teams and Team Mechanics" and provides a rationale for utilizing teams in performing HAZOP analyses. It also defines what is necessary for a team to be effective, qualifications of team members, consensus, team process checks, problem solving, and logistical issues concerning meetings.

Chapter 9 discusses job hazard analysis and OSHA regulations and how they effect and affect risks in work environments. Chapter 10 is titled "Hazard Communication Based on CFR 910.1200" and covers a typical automotive hazard communication program. Specifically, it addresses the individuals involved, their responsibilities, appropriate training, and the importance of safe use instructions, chemical materials lists, and material safety data sheets.

Appendix A provides sample checklists for devising a safety plan and a facility location plan and guidelines of the Australian Health Administration. Appendix B details a HAZOP project.

## Bibliography

Callaghan, B. and T. Walker. (2001). *Reducing Risks: Protecting People: Decision-Making Process*. Norwich, U.K.: Crown Publications.
http://www.hse.gov.uk/risk/theory/r2p2.pdf

# Acknowledgments

No single individual is ever capable enough to undertake any topic and develop it into a book form so that others may benefit. Everyone depends on many individuals who either have contributed to that topic before commencement of the work or during the writing. This work is no different.

Even though I have been in the field of quality for over 30 years, I have always valued the contributions of others in the areas of their expertise and certainly in contributing to the development of my own thoughts and suggestions for the topics that I write about. Over the years, I have come to appreciate many colleagues for their suggestions and recommendations on issues dealing with reliability, risk, and quality in general. To mention them all is impractical. However, there are some who stand out both because of their suggestions and also for their faith in me to complete this project. I am grateful indeed to several individuals.

C. Wong from APRC in Singapore has relentlessly been one of my strongest friends and a generator of ideas for writing technical books. He was instrumental in persuading me of the need of this book, and I finally took the project on.

Ka Wong showed dedication and thoroughness in drawing the figures in this book patiently and without complaining even after they had to be redrawn when I lost them all in a computer error.

L. Lamberson, PhD, and professor of mechanical engineering at Western Michigan University provided help with some technical issues, particularly the FTA and ETA techniques, and helped me address them in simple terms.

R. Kapur, PhD, director of quality at Tompkins Products posed thought-provoking questions about ALARP during the writing of this book.

Ken Wolf of Ford Motor Company made many detailed suggestions for handling FTA and ETA as they relate to traditional FMEA.

C. Michalakis, the GM-UAW safety coordinator, provided insight about safety and hazards communications. His help in explaining some of the first-line issues made this book more realistic and practical.

My son, C. Stamatis, prepared the graphs on a computer and helped with all the logistical computer issues that arose during the writing.

My other son, S. Stamatis, PhD, spent countless hours debating and discussing chemical hazards and their impacts on organizations and society at large.

Ian Sutton permitted me to use the interconnectivity and node figures from http://www.stb07.com/process-safety-management/process-hazards-analysis.html and http://www.stb07.com/process-safety-management/hazop.html

I thank the State of New South Wales through the Department of Planning for giving me permission to use Figure 1, 2, 3, and 4 January 2008 and 2011 Pg. vi, 7, 25-31 and 33. Hazardous Industry Planning Advisory Paper No 8. (In this book they are Figures I.1,5.3 and Appendix B). HAZOP Guidelines are from www.planning.nsw.gov.au

I thank Elsevier Publishing for granting permission through the Copyright Clearance Center for using material from Chapter 3 of Sutton's 2010 book titled *Process Risk and Reliability Management*.

Thanks also to the editors for a superb job on the layout and improvements to the original manuscript. Your efforts made this book more readable and certainly more functional to follow.

I thank all my clients and friends who provided me with insights many times in the application of risk analysis in the area of hazards, including safety and environmental issues.

Finally, the biggest thank you goes to my chief editor and critic—my wife, Carla. She has been very supportive during the entire project, pulling me out of lethargic moods and encouraging me to continue writing. Without her, this book would never have been finished.

# Author

**Dean H. Stamatis, PhD, ASQC Fellow, CQE, CMfgE, MSSBB, ISO 9000 Lead Assessor (graduate)**, is the president of Contemporary Consultants Co. in Southgate, Michigan. He is a specialist in management consulting, organizational development, and quality science. He has taught project management, operations management, logistics, mathematical modeling, economics, management, and statistics at both graduate and undergraduate levels at Central Michigan University, University of Michigan, ANHUI University (Bengbu, China), University of Phoenix, and Florida Institute of Technology.

With over 30 years of experience in management, quality training, and consulting, Dr. Stamatis has served numerous private sector industries, including steel, automotive, general manufacturing, tooling, electronics, plastics, food, maritime, defense, pharmaceutical, chemical, printing, healthcare, and medical device industries.

He has consulted for such companies as Ford Motor Co., Federal Mogul, GKN, Siemens, Bosch, SunMicrosystems, Hewlett Packard, GM Hydromatic, Motorola, IBM, Dell, Texas Instruments, Sandoz, Dawn Foods, Dow Corning Wright, British Petroleum, Bronx North Central Hospital, Mill Print, St. Claire Hospital, Tokheim, Jabill, Koyoto, SONY, ICM/Krebsoge, Progressive Insurance, B. F. Goodrich, and ORMET, to name just a few.

Dr. Stamatis has created, presented, and implemented quality programs with a focus on total quality management, statistical process control (both normal and short run), design of experiments (both classical and Taguchi), Six Sigma (DMAIC and DFSS), quality function deployment, failure mode and effects analysis (FMEA), value engineering, supplier certification, audits, reliability and maintainability, cost of quality, quality planning, ISO 9000, QS-9000, ISO/TS 16949, and TE 9000 series. He has created, presented, and implemented programs on project management, strategic planning, teams, self-directed teams, facilitation, leadership, benchmarking, and customer service.

Dr. Stamatis is a certified quality engineer through the American Society of Quality Control, a certified manufacturing engineer through the Society of Manufacturing Engineers, a certified master black belt through IABLS, and is a graduate of BSI's ISO 9000 lead assessor training program.

He has written over 70 articles, presented many speeches, and participated in national and international conferences on quality. He is a contributing author to several books and the sole author of 42 books. His consulting extends across the United States, South East Asia, Japan, China, India, Australia, Africa, and Europe. In addition, he has performed more than 100 automotive-related audits, 25 pre-assessment ISO 9000 audits, and helped several companies attain certification, including Rockwell International's Switching Division (ISO 9001), Transamerica Leasing (ISO 9002), and Detroit Electro Plate (QS-9000).

Dr. Stamatis earned a BS/BA in marketing from Wayne State University, a master's from Central Michigan University, and a PhD in instructional technology, business, and statistics from Wayne State University. He is an active member of the Detroit Engineering Society and the American Society for Training and Development, an executive member of the American Marketing Association, a member of the American Research Association, and a fellow of the American Society for Quality.

# Introduction

An important element of any system for the prevention of major accidents is conducting a hazard and operability study (HAZOP) at the detail design stage of a plant in general and operating and safety control systems in particular. HAZOP analysis seeks to minimize and/or eliminate if possible the effect of an atypical situation on an operation or process. It does that by ensuring that control and other safety systems such as functional safety measures, or safe shutdown devices, are in place and work with a high level of reliability to ensure a safe outcome from a situation that could have become a major accident.

Like all other prevention tools and/or methods HAZOP is used to identify potential hazards and operational problems arising from plant design and human error. The technique is applied during final design of the process and plant items before commencement of construction. However, in the last 5 or so years, HAZOPs have also yielded financial benefits by minimizing the time and money spent in installing additional control and safety systems. This approach has been very fruitful since it may identify additional controls before construction begins and may reveal needs that otherwise would have become evident only after plant commissioning. Where operability is a concern, benefits are accumulated by implementing the remedial recommendations to operability issues identified by a HAZOP during the design stage.

A HAZOP analysis describes a hazard as the potential for harm arising from an intrinsic property (inherent to a task) or disposition of something that may cause detriment. It defines risk as the chance that someone or something that is valued will be adversely affected by some loss of value. This view is of particular interest when people are components of a risk. A risk that involves people, like any other risk, must be evaluated and appropriate control measures introduced to address the identified risk. This is of profound importance because by controlling the risk we imply that it exists and may happen, but we will have a system in place to contain the ramifications of the risk if it occurs. The focus is on prevention and control rather than on detection after the fact. This focus must be realized up front because *risk* in this contest means *possibility of danger* rather than actual danger. The possibility of danger exists because we have no guarantee of absolute safety and zero hazard in any activity.

To compensate for this possibility, we use the principles of (1) so far as is reasonably practicable (SFAIRP), (2) as low as reasonably practicable (ALARP), and (3) as low as reasonably achievable (ALARA). We use these principles—some call them methodologies—to avoid a potential hazard and to ensure within reason that preventive and protective actions are appropriate and applicable for the identified risk.

Why do we need and perform an analysis to identify and understand risks? Because all risks are regarded as tolerable; however, *tolerable* does not mean *acceptable*. A tolerable risk is a willingness by society and/or a specific organization to live with a particular risk to (1) secure certain benefits and (2) be confident that the risk is worth taking and is being properly controlled. However, this does not imply that a risk will be acceptable to everyone, i.e., that all parties would agree without reservation to take the risk or have it imposed on them. Different individuals have different tolerances of accepting risk and that is what makes the study of risks interesting.

In our modern world, we have a tendency to evaluate benefits and risks in any industrial activity against any undesirable side effects such as oil spills, environmental pollution, and so on. This tendency is more specific and true for risks that

- May lead to catastrophic consequences
- May have irreversible consequences such as the release of genetically modified organisms
- May create inequalities because they affect some people more than others such as risks arising from the location of a chemical plant or waste disposal facility
- May pose a threat to future generations, for example, toxic wastes.

The results of this analysis in modern organizations is that many industries have less discretion in areas where they previously had considerable freedom to decide which course of action to adopt and pursue. Examples are plans for modifying plants within their own boundaries; selections of raw materials and processes; and how precautionary actions should be identified and to what extent.

For these reasons, preliminary HAZOPs are performed. A typical approach is shown in Figure I.1. A preliminary HAZOP addresses major key components of a system such as emergency plans, fire precautions, construction issues, and safety concerns. The intent of a preliminary HAZOP is to evaluate feasibility as early as possible and circumvent catastrophic problems down the road. More details are detail is given in Chapter 4. After a preliminary HAZOP has been identified, the selection of specificity begins. The flow of this process is shown in Figure I.2. By contrast, node selection is shown in Figure I.3.

Many systems have been developed for informing and reaching decisions, and those pertinent to health and safety are more descriptive than others. However, all of them follow six stages (Callaghan and Walker 2001):

- Stage 1: Deciding whether an issue concerns health, safety and environment, or organizational improvement
- Stage 2: Defining and characterizing the issue

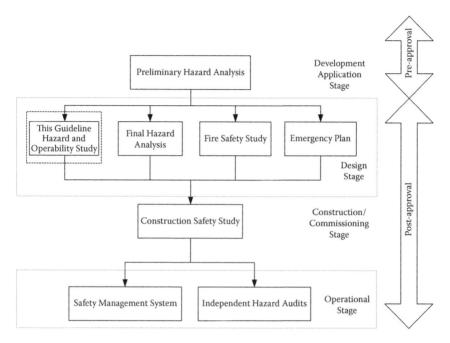

**FIGURE I.1**
Preliminary HAZOP. (*Source:* HIPAP 8, New South Wales, Australia. With permission.)

- Stage 3: Examining the options available for addressing the issue and their merits
- Stage 4: Adopting a course of action for addressing the issue efficiently and in good time, informed by the findings of the second and third points above and in the expectation that as far as possible it will be supported by stakeholders
- Stage 5: Implementing the decisions
- Stage 6: Evaluating the effectiveness of actions taken and revisiting the decisions and their implementation if necessary

Four points related to these six stages are worth emphasizing. First, one may get the impression that each stage is distinct and independent. In fact, in reality, the boundaries between them are not clear-cut and often overlap. Generally, we collect information or perspectives while progressing from one stage to another. This process forces us to look and evaluate earlier stages; thereby, we continually improve the process in an iterative and dynamic mode.

Second, in all stages consensus is of primary concern and therefore designated team members must actively participate in discussions and the decision-making process. It is possible that consensus may not be reached,

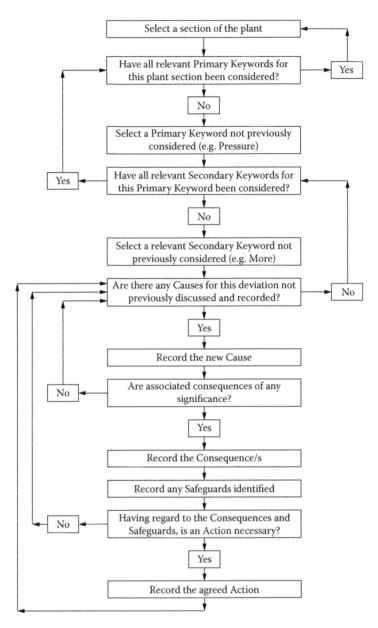

**FIGURE I.2**
Selection of HAZOP process.

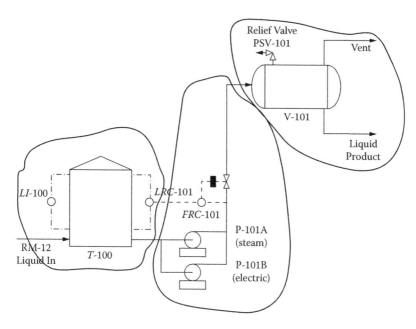

**FIGURE I.3**

Node selection. (*Source:* Ian Sutton, http://www.stb07.com/process-safety-management/hazop.html. With permission.)

in which case multiple meetings must be held to allow the team to reach an amicable resolution.

Third, it is imperative to recognize a team has no organized or standardized format to follow at any stage because the process under evaluation is not fixed at this point. It is dynamic and subject to change based on information gained formally and informally.

Finally, at every stage, the team must be careful to incorporate lessons learned from previous applications with a major caveat: be careful of past actions because regulations and/or circumstances may have imposed changes that make earlier actions inappropriate or inapplicable to the new process. Furthermore, some circumstances may require that actions are taken quickly due to emergencies or other unforeseen events. Obviously, this occurs because risk always involves uncertainty and any process under consideration may be more complex than anticipated because sufficient and applicable data are not available.

Uncertainty is a state of knowledge in which, although the factors influencing the issue are identified, the likelihood of any adverse effects or the effects cannot be described precisely. Uncertainty has many manifestations, and they affect the approach to its handling. Key items of concern are as follows (Callaghan and Walker 2001):

- Knowledge uncertainty: This arises when knowledge is represented by data based on limited statistics or subject to random errors in experiments. In that case, confidence limits must be addressed in a sensitivity analysis. This approach will lead the team to information relating to the importance of different sources of uncertainty that can then be used to prioritize actions or pursue more study of the process.
- Modeling uncertainty: This concerns the validity of the choice of representation in mathematical terms or via a process mapping approach. For each choice, the risks will be identified and the definitions of value and nonvalue will be much easier to identify. An example is the growth of a crack in the wall of a pressure vessel. The model would
  - Formulate a mathematical equation showing how growth rate is affected by factors such as the material properties and the stress history of the vessel.
  - Provide predictions of the time and nature of the failure. Testing to failure is important here.
  - Communicate and inform intervention strategies such as material specifications, in-service monitoring, and mitigation measures.

  In modeling for uncertainty, just like other modeling endeavors, it is important to identify and select factors that the team knows will affect the risk and/or review the literature identifying factors based on theoretical grounds. Obviously, a variety of mathematical models may be utilized, but they all require open minds and a set of agreed assumptions. The integrity and the soundness of the selected factors for analysis are fundamentally based on team discussions and alternative hypotheses for testing the various scenarios. We must always be cognizant that no one may have an idea how to model some issues or the expected consequences of the uncertainty. In this case, expert *judgment*, also known as *management bias*, prevails. The bias may be real or perceived. The fact that it is subjective and dictated makes it a *bias* by definition.
- Limited predictability or unpredictability: There are limits to the predictability of phenomena when the outcomes are very sensitive to the assumed initial conditions. Systems that begin in a nominal (target) state do not end up in the same final state. Any inaccuracy in determining the actual initial state will limit our ability to predict the future, and in some cases, the system behavior will become unpredictable. In looking at options according to www.HSE.gov.uk, we would be particularly interested in examining:
  - *Possible good practice* for addressing the hazards identified and evaluating whether it is relevant, efficient, effective, and sufficient. If specific good practice is not available, we would examine the merits of good practice that apply in comparable circumstances via *surrogate data* if we believe that the information

is directly transferable or can be suitably modified to address the hazard. As we gain knowledge, the *surrogate data* must be replaced with actual data, and the analysis must be redone so that the new data may be validated.

- *Possible constraints attached to a particular option* must be addressed as early as possible. Typical constraints are usually technical issues such as feasibility, legal regulations, and possible political ramifications.

- *Any adverse consequences associated with a particular option* must be identified, discussed, and resolved as early as possible. Sometimes when we resolve an issue, we create some other issue as a result of our action. When this happens, we must recognize that we have not resolved the first issue; in reality, we created a new issue.

- *How much uncertainty is attached to the issue under consideration* and as a consequence the precautionary approach is of importance as well. If we know a limit of acceptable uncertainty, it should be adopted to ensure that the outcome of a team's decisions is in congruence with the precautionary principle that we have set up in defining our risk. One more important characteristic is that in cases where the benefits cannot justify the risks, it is imperative to abandon the activity, process, or practice that contributes to the hazard.

- *How far certain options should be constrained* is another inherent team concern. It is an important consideration because it defines the limits of the scenario at hand. Of course, this limit should be part of the scope defined initially. For example, when a team is considering improving the health and safety for a steel company, in specific terms they start analyzing the appropriateness of investments of their steel pourer's operators. Obviously the team has gone beyond the scope of the task. Investments are important but not part of a safety risk analysis for a steel company.

- *How far the options succeed in improving (or at least maintaining) standards* is yet one more requirement of the team in determining level of success.

- *The costs and benefits* are also important. Every aspect of the recommendations requires a separate document explaining the cost and especially what is required to implement the specific recommended action in relation to the risk identified.

- *What is the most appropriate regulatory instrument* is perhaps the most overlooked requirement in the evaluation process. This should be investigated by the team as it represents the regulations that apply to the particular task. By identifying the regulations, the team and management have a better chance to manage a risk if it materializes later.

After we have gathered all the necessary information, we are ready for the appropriate review intended to determine the most appropriate option for managing the risks. The key to success depends to a large extent on ensuring as far as possible that interested parties are content with the process for reaching decisions and, hopefully also with the decisions. For example, they should be satisfied with (1) the way uncertainty has been addressed and the plausibility of the assumptions made; and (2) how other relevant factors such as economic, technological, and political considerations have been integrated in the decision-making process.

Meeting these conditions is not always easy, particularly when parties have opposing opinions based on differences in fundamental values or concentrate on a single issue. Nevertheless, we tackle the first condition by

- Finding and focusing on the uncertainties that matter
- Explaining why a particular method of estimating risks was chosen in preference to others
- Being open to the science, assumptions, and other critical inputs that contributed to the value or judgment obtained from the risk assessment exercise

Addressing the second condition above (determining how economic, technological, and political considerations were integrated in the decision-making process) is more difficult. Success lies in adopting decisions that most accurately reflect the ethical and value preferences of society at large; on what risks are unacceptable, tolerable, or broadly acceptable; and how successful we have been in involving stakeholders in the decision-making process. At times, taking account of uncertainty and the need to adopt a precautionary approach may require more focus on the consequences of harm from a hazard than on the likelihood that the hazard will occur.

When we have reached a decision on the degree to which a risk should be controlled, we must decide how the decision can be implemented in practice using the regulatory tools at our disposal. The *how* may be by recommending new legislation, inviting new guidance, or taking enforcement action on the current regulations and expectations. The responsibility for measures for controlling a risk will usually fall on the person who creates it or is in a position to prevent or minimize it. When constructing a regulatory tool, we should apply an approach that is comparable with the regulation in a way that it:

- Is exposed to the checks and balances inherent in government regulations for dealing with occupational health and safety matters, thus ensuring compliance with fundamental principles, i.e., that the strategy and targets are not compromised and that societal concerns are considered.

- Involves consulting the stakeholders and requires communicating the outcome to them effectively.
- Fits into the context of legal requirements and requires those who must introduce measures for managing risks to
  - Enlist the cooperation and involvement of those affected and those able to assist, such as safety representatives, by pointing out that the activity is crucial for the proper management of health and safety. For example, the involvement of safety representatives in health and safety management can help the responsible personnel (process owners) to fulfill their legal obligations and achieve high standards of health and safety. Moreover, employers are unlikely to achieve the proper control of risks in their workplace without the help of their employees.
  - Introduce procedures that foster a culture disposing all involved parties to deliver their best. In the workplace, this may mean getting a commitment at every level of the organization to adopt high health and safety standards and work toward them. It also calls for the establishment of well-considered and articulated safety policies that properly define and allocate responsibilities and organizational arrangements intended to ensure control and promote cooperation, communication, and competence.
  - Have a plan for taking action by looking ahead and setting priorities for ensuring that risks requiring the most attention are tackled first, based on a risk assessment that may be required by specific legislation.
  - Set up a system for monitoring and evaluating progress by identifying potential indicators for evaluating how far the control measures introduced have been successful in addressing the problem.
  - Comply with a well-established hierarchy of measures for the prevention of risks, e.g., eliminating risks, combating risks at their sources, generally applying sound engineering practices such as inherently safer design, and applying collective protective measures rather than individual protective measures.

Individual employees also have duties to

- Take reasonable care of their own health and safety and the health and safety of other persons who may be affected by acts or omissions at work.
- Cooperate with their employers as necessary to enable employers to comply with their statutory health and safety responsibilities.

Finally, our process for ensuring that risks are properly managed would not be complete without procedures to review our decisions after a suitable interval to establish:

- Whether the actions taken to ensure that the risks are adequately controlled produced the intended result.
- Whether decisions previously reached need to be modified and, if so, how; for example, because levels of protection that were considered to be good practices may no longer be regarded as such as a result of new knowledge, advances in technology, or changes in the level of societal concerns.
- The appropriateness of the information gathered in the first two stages of the decision-making process to assist decisions for action, i.e., the methodologies used for the risk assessment and the cost–benefit analysis (if prepared) or assumptions made.
- Whether improved knowledge and data would have led to better decisions.
- What lessons could be learned to guide future regulatory decisions, improve the decision-making process, and create greater trust among regulators, operators, and those affected by or having an interest in the risk problem.

An appropriate evaluation should be based on *equity-based* criteria that start with the premise that all individuals have unconditional rights to certain levels of protection. This leads to standards, applicable to all, held to be usually acceptable in normal life or based on some other premise held to establish an expectation of protection. In practice, this often converts into fixing a limit to represent the maximum level of risk above which no individual can be exposed. If the risk estimate derived from the risk assessment is above the limit and further control measures cannot be introduced to reduce the risk, the risk is held to be unacceptable whatever the benefits.

We also face *utility-based* criteria that apply to the comparison of the incremental benefits of the measures to prevent the risk of injury or detriment and the costs of the measures. In other words, utility-based criteria compare in monetary terms the relevant benefits (e.g., lives saved, equipment life-years extended) obtained by the adoption of a particular risk prevention measure with the net cost of introducing it. These criteria require that a particular balance be struck between the two factors. This balance can be deliberately skewed toward benefits by ensuring a gross disproportion between the costs and benefits.

Yet another approach to evaluation is a *technology-based* criterion that essentially reflects the idea that a satisfactory level of risk prevention is attained when state-of-the-art control measures (technological, managerial, and organizational) are employed to control risks, whatever the circumstances.

## References

Callaghan, B. and T. Walker. (2001). *Reducing Risks: Protecting People: Decision-Making Process*. Norwich, U.K.: Crown Publications.

HIPAP 8. (January 2011). *HAZOP Guidelines*. Sydney: State of New South Wales Department of Planning.

http://www.hse.gov.uk/risk/theory/r2p2.pdf

## Selected Bibliography

Cowan, N. (2005). *Risk Analysis and Evaluation,* 2nd ed., Kent, U.K.: Institute of Financial Services.

# 1

## Risk

### General Definition

Risk is everywhere, in everything we do. In all areas of life, risk is something that we must manage, whether we direct a major organization or simply cross a road. When describing risk, however, it is convenient to consider that risk practitioners operate in specific practice areas.

In general terms, *risk* is the potential that a chosen action or activity, including the choice of inaction, will lead to a loss or undesirable outcome. This, of course, implies that a choice that will influence the outcome exists or existed. Potential losses may also be called risks. Almost any human endeavor carries some risk, but some activities are much riskier than others.

The ISO 31000 (2009)/ISO Guide 73:2002 definition of risk (see Note 1 at the end of this chapter) is the "effect of uncertainty on objectives." In this definition, uncertainties include events (that may or may not happen) and uncertainties caused by ambiguity or a lack of information. It also includes both negative and positive impacts on objectives. On the other hand, although many definitions of risk exist in common usage, the predominant one is the ISO definition developed by an international committee representing over 30 countries and based on the inputs of several thousand subject matter experts (SMEs).

Although many of us use the word *risk* with different connotations, the *Oxford English Dictionary* cites the earliest use of the word in English (spelled as *risque*) in 1621 and the *risk* spelling from 1655. The dictionary defines *risk* as (exposure to) the possibility of loss, injury, or other adverse or unwelcome circumstance; a chance or situation involving such a possibility (Dodge 2003).

We have come a long way in our understanding of the modern meaning of *risk*. In fact, for the sociologist Luhmann (1996), *risk* is a neologism that appeared during the transition from traditional to modern society. Franklin (2001 p. 274) informs us that, "in the Middle Ages, the term *risicum* was used in highly specific contexts, above all sea trade and its ensuing legal problems of loss and damage." In the common languages of the 16th century, *rischio* and *riezgo* were used. The term was introduced to continental Europe through interaction with Middle Eastern and North African Arab traders. In

1

the English language, *risk* appeared only in the 17th century, and "seemed to be imported from continental Europe." When the terminology of *risk* took ground, it replaced the older notion of "good and bad fortune." Luhmann (1996), in trying to explain this transition, wrote, "Perhaps, this was simply a loss of plausibility of the old rhetorics of *Fortuna* as an allegorical figure of religious content and of *prudentia* as a (noble) virtue in the emerging commercial society."

As the studies of risk proliferated in many sectors of the economy, *scenario analysis* became a primary methodology for dealing with risk. Scenario analysis matured during Cold War confrontations between the United States and the Soviet Union. It became widespread in insurance circles in the 1970s when major oil tanker disasters such as the *Exxon Valdez* incident off the coast of Alaska forced a more comprehensive examination of marine risk issues. The scientific approach to risk entered finance in the 1960s with the advent of the capital asset pricing model and became increasingly important in the 1980s when financial derivatives proliferated. It reached general business in the 1990s when the power of personal computing allowed for widespread data collection and number crunching. Governments use scenario analysis, for example, to set standards for environmental regulation. The U.S. Environmental Protection Agency utilizes pathway analysis.

## Other Definitions

The many inconsistent and ambiguous meanings attached to *risk* led to widespread confusion and the development of various approaches to risk management in different fields (Hubbard 2009). For example, Jones (2005) sees risk as relating to the probability of uncertain future events and describes the probable frequency and probable magnitude of future loss. In computer science, this definition is used by The Open Group (see Note 2).

OHSAS (Occupational Health & Safety Advisory Services 18001:2007) defines risk as the product of the probability of a hazard resulting in an adverse event times the severity of the event. In information security, risk is defined as "the potential that a given threat will exploit vulnerabilities of an asset or group of assets and thereby cause harm" to an organization (ISO/IEC 27005:2008).

Financial risk is often defined as the unexpected variability or volatility of returns and thus includes worse-than-expected and better-than-expected returns. References to negative risk should be understood as applying to positive impacts or opportunities (losses *or* gains) unless the context precludes this interpretation. The related *threat* and *hazard* terms are often used to mean an event that could cause harm.

## Economic Risks

Economic risks can be manifested in lower returns or higher expenditures than expected. The causes can be many, for instance, a hike in the prices of raw materials, the lapsing of deadlines for construction of a new operating facility, disruptions in a production process, emergence of a serious competitor in a market, the loss of key personnel, a change of political regime, or natural disaster (Galasyuk and Galasyuk 2007). Reference class forecasting was developed to eliminate or reduce economic risk (Flyvbjerg 2008).

It is worth noting that from a societal standpoint losses are much more lucrative than gains because governmental bodies will do anything required, according to recent research, to avoid losing or resorting to an inferior position (Nichols 2000 p. 4).

## Health Risks

Risks to personal health may be reduced by primary prevention actions that decrease early causes of illness or by secondary prevention actions after measured clinical signs or symptoms are recognized as risk factors. Tertiary prevention reduces the negative impact of an already established disease by restoring function and reducing disease-related complications.

Ethical medical practice requires careful discussion of risk factors with individual patients to obtain informed consent for secondary and tertiary prevention efforts. Public health efforts in primary prevention require education of a population at risk. In both cases, careful communication about risk factors, likely outcomes, and certainties must distinguish causal events that must be decreased from associated events that may be merely consequences rather than causes. In epidemiology, Rychetnik et al. (2004) reported that the *lifetime risk* of an effect is the *cumulative incidence*, also called *incidence proportion*, over an entire lifetime (see Note 3).

## Health, Safety, and Environment (HSE) Risks

Health, safety, and environment (HSE) risks constitute separate practice areas, but they are often linked. The reason typically concerns organizational structures, but all three disciplines are linked strongly. One of the strongest links is that a single risk event may have impacts in all three areas, albeit over differing timescales. For example, the uncontrolled release of radiation or a toxic chemical may have immediate short-term safety consequences, produce more protracted health impacts, and create longer-term environmental impacts. Events such as the Chernobyl nuclear failure in April 1986 caused immediate deaths, later deaths from cancers, and left a lasting environmental impact that led to birth defects, impacts on wildlife, and other negative results. The catastrophic 2011 Tohoku earthquake and tsunami in Japan caused similar devastation.

### Information Technology (IT) and Information Security Risks

Information technology risk, also known as IT risk or IT-related risk, arises from the use of information technology. This relatively new term developed from an increasing awareness that information security is simply one facet of a multitude of risks related to IT and the real-world processes it supports. The increasing dependencies of modern society on information and computer networks in the private and public sectors, including the military (Cortada 2003 p. 512, 2005, 2007 p. 496), led to new terms like *IT risk* and *cyber warfare*.

*Information security* involves the protection of information and information systems from unauthorized access, use, disclosure, disruption, modification, perusal, inspection, recording, and destruction (44 USC §3354(b)(1)). Information security grew out of practices and procedures for ensuring computer security and developed into *information assurance* (IA), defined as the practice of managing risks related to the use, processing, storage, and transmission of information or data and the systems and processes used for those purposes. While focused dominantly on information in digital form, the full range of IA encompasses digital, analog, and physical data. IA is interdisciplinary and draws from multiple fields, including accounting, fraud examination, forensic science, management science, systems engineering, security engineering, criminology, and computer science.

IT risk is narrowly focused on computer security, while *information security* extends to risks related to other forms of information (paper, microfilm). IA risks include those related to the consistency of the data stored in IT systems, data stored on other means, and the relevant business consequences.

### Insurance Risks

Insurance is a risk treatment option that involves risk sharing. It can be considered a form of contingent capital and is akin to purchasing an option at a small premium to acquire protection from a potential large loss. Insurance companies assume pools of risks, usually for profit, and may specialize in areas such as market risk, credit risk, operational risk, interest rate risk, mortality risk, and longevity risk (Carson et al. 2008).

### Business and Management Risks

Means of assessing risk vary widely among professions. For example, a doctor manages medical risk, while a civil engineer manages risk of structural failure. A professional code of ethics is usually focused on risk assessment and mitigation by professionals on behalf of clients, the public, society, or life in general. Incidental and inherent risks exist in all workplaces. Incidental risks occur naturally in the course of business but are not part of the business core. Inherent risks have negative effects on the operating profit of a business.

## Human Services Risks

Important ethical and political issues arise when humans are seen or treated as risks or when the risk decisions are made by people who use human services may have impacts on the services. The experiences of many people who rely on human services for support is that risk is often used as a reason to prevent them from gaining further independence or fully accessing the community, and that the services are often unnecessarily risk averse (Neil et al. 2008). "People's autonomy used to be compromised by institution walls, now it's too often our risk management practices" (Sanderson and Lewis 2011).

## High Reliability Organizations (HROs)

HROs are organizations that have succeeded in avoiding catastrophes in environments where normal accidents can be expected due to risk factors and operational complexities. Most studies of HROs involve nuclear aircraft carriers, air traffic control, aerospace explorations, and nuclear power stations. Such organizations share the ability to consistently operate safely in complex, interconnected environments, where a single failure of one component could lead to catastrophe. Essentially, they appear to operate in spite of an enormous range of risks.

Some of these industries manage risk in a highly quantified way. In the nuclear power and aircraft industries, the possible failure of a complex series of engineered systems could result in highly undesirable outcomes. The usual measure of risk for a class of events is calculated as:

$$R = \text{probability of event} \times \text{severity of consequence}$$

Given the risk for an event, we can determine the total risk by calculating the products of the individual class risks. For example, in the nuclear industry, consequence is often measured in terms of off-site radiological release, often banded into five or six decade-wide bands.

The risks are evaluated using fault tree and event tree techniques (see Chapters 6 and 7). Where these risks are low, they are normally considered *broadly acceptable*. A higher level of risk (typically 10 to 100 times what is considered broadly acceptable) must be justified against the costs of reducing the risk further and the possible benefits that make it tolerable. These risks are described as *tolerable if as low as reasonably practicable (ALARP)*. Risks beyond this level are classified as *intolerable*.

The broadly acceptable level of risk has been considered by regulatory bodies in various countries. The rationale for such acceptability was demonstrated by F. R. Farmer who showed that certain risk is acceptable to individuals even though the activity considered presents definable risks. He demonstrated this by hill walking and similar activities. The results of his findings were presented as the now famous Farmer curve of acceptable probability of an event versus its consequence (Ayyub 2003 p. 101).

The technique as a whole is usually known as probabilistic risk assessment (PRA) or probabilistic safety assessment (PSA). The WASH-1400 report, also known as *The Reactor Safety Study*, is an example of this practice. The report was produced in 1975 for the Nuclear Regulatory Commission by a committee of specialists. However due to heavy criticism that raised many questions regarding its assumptions, methodology, calculations, peer review procedures, and objectivity, the report was declared obsolete and replaced by the *State-of-the-Art Reactor Consequence Analyses*.

## Security Risks

Security risk management involves protection of assets from harm caused by deliberate acts. A more detailed definition by Talbot and Jakeman (2009) is: "A security risk is any event that could result in the compromise of organizational assets, the unauthorized use, loss, damage, disclosure or modification of organizational assets for the profit, personal interest or political interests of individuals, groups or other entities constitutes a compromise of the asset, and includes the risk of harm to people. Compromise of organizational assets may adversely affect the enterprise, its business units and their clients. As such, consideration of security risk is a vital component of risk management. Table 1.1 lists the risk-related sections from ISO/IEC Guide 73:2002 and 2009.

## Societal Risks

In a peer-reviewed study of risk in public works projects in 20 nations on 5 continents, Flyvbjerg et al. (2002, 2005) documented high risks for such ventures based on both costs and demands. Actual costs of projects were typically higher than estimated costs; cost overruns of 50% were common, overruns above 100% were not uncommon. Actual demand was often lower than estimated; demand shortfalls of 25% were common, 50% not uncommon.

**TABLE 1.1**

ISO/IEC 27001 Clauses Related to Risk

| Topic | 2002 Clause | 2009 Clause |
|-------|-------------|-------------|
| 3.9 | Residual risk | 3.8.1.6 |
| 3.10 | Risk acceptance | 3.7.1.6 |
| 3.11 | Risk analysis | 3.6.1 |
| 3.12 | Risk assessment | 3.4.1 |
| 3.13 | Risk evaluation | 3.7.1 |
| 3.14 | Risk management | 2.1; 3.1 |
| 3.15 | Risk treatment | 3.8.1 |

Due to such costs and demand risks, cost–benefit analyses of public works projects have proven highly uncertain. The main causes of cost and demand risks were found to be optimism bias and strategic misrepresentation. Measures identified to mitigate such risks include better governance through incentive alignment and the use of reference class forecasting (Flyvbjerg 2004).

## Human Factors Risks

One of the growing areas of focus in risk management is the human factors field in which behavioral and organizational psychology underscores our understanding of risk-based decision making. This field considers questions such as how we make risk-based decisions and why we are irrationally more scared of sharks and terrorists than we are of motor vehicles and medications. In decision theory, regret (and anticipation of regret) can play a significant part in decision making, distinct from risk aversion (preferring the status quo to prevent being worse off); see Note 4.

Tversky and Kahneman (1981) define framing as a fundamental problem with all forms of risk assessment (see Note 5). In particular, because of bounded rationality (the human brain takes mental shortcuts when it becomes overloaded), the risk of extreme events is discounted because the probability is too low to evaluate intuitively. As an example, among the leading causes of death are vehicle accidents caused by drunk driving— partly because drivers frame the problem by minimizing or ignoring the risks of serious or fatal accidents.

Another example is ignoring an extremely disturbing event (such as a hijacking) in a risk analysis despite the fact it has occurred and has a nonzero probability. An event that everyone agrees is inevitable may be ruled out of analysis due to greed or an unwillingness to admit that it may be inevitable. These human tendencies based on wishful thinking that allow for error often affect even the most rigorous applications of the scientific method and are major concerns of the philosophy of science.

All decision making under uncertainty must consider cognitive bias, cultural bias, and notational bias. No group of people assessing risk is immune to *groupthink*. A consequence of a groupthink decision may be an acceptance of obviously wrong answers simply because it is socially painful to disagree where there is conflict of interest.

Framing involves other information that affects the outcome of a risky decision. The right prefrontal cortex has been shown to take a more global perspective (Schatz et al. 2004) while greater left prefrontal activity relates to local or focal processing (Schatz et al. 2004a).

Based on Drake's (2004) "Theory of Leaky Modules," McElroy and Seta (2004) proposed that they could predictably alter the framing effect by the selective manipulation of regional prefrontal activity with finger tapping or monaural listening. The result was as expected. Rightward tapping or listening had the effect of narrowing attention such that the frame was ignored. This is a practical way of manipulating regional cortical activation to affect risky decisions, especially because directed tapping or listening is easily done.

## Risk Assessment and Analysis

Since planned actions are subject to large cost and benefit risks, proper risk assessment and risk management strategies are crucial for making such actions successful (Flyvbjerg 2006; Note 6). Since risk assessment and management are essential components of security management, they are tightly related. Security assessment methodologies like the CCTA risk analysis and management method (CRAMM) contain risk assessment modules in the first steps of the methodology (Note 7). Techniques like the method for harmonized analysis of risk (MEHARI) evolved to become security assessment methodologies (Note 8). The ISO standard on risk management (principles and guidelines on implementation) was published as ISO 31000 on November 13, 2009.

## Quantitative Analysis

Because *risk* carries so many meanings, several formal methods are used to assess or measure it. Some of the quantitative definitions of risk are grounded in statistics theory and lead naturally to statistical estimates, and some are more subjective. For example, human decision making is a critical factor in many situations. Even when statistical estimates are available, in many cases risk is associated with rare types of failures and data may be sparse.

Often the probability of a negative event is estimated by using the frequency of past similar events (surrogate data) or by event tree methods, but probabilities for rare failures may be difficult to estimate if an event tree cannot be formulated. This makes risk assessment difficult in hazardous industries, for example, nuclear energy, where the frequency of failures is low but the harmful consequences of failure are numerous and severe.

Statistical methods may also require the use of a cost function, which in turn may require the calculation of the cost of loss of a human life. This is a difficult problem. One approach is to ask what people are willing to pay to insure against death (Landsburg 2003) or radiological release (such as large quantities of radioactive iodine). Because the answers depend strongly on the circumstances, this approach is clearly not effective. In statistics, the notion of risk is often modeled as the expected value of an undesirable outcome. This combines the probabilities of various possible events and some assessment of the corresponding harm into a single value. In statistical decision theory, the risk function is defined as the expected value of a given loss as a function of the decision rule used to make decisions in the face of uncertainty.

In the fields of finance and economics, risk is referred to as *expected utility*. The simplest case is a binary possibility of *accident* or *no accident*. The associated formula for calculating risk is then:

$$\text{Risk}_{\text{of event}} = \text{probability of an accident occurring} \times \text{expected loss if accident occurs}$$

For example, if performing activity X has a probability of 0.01 of suffering an accident A with a loss of 1000, total risk is a loss of 10 = (0.01 × 1000). Obviously, situations may be far more complex than the simple binary possibility case. In a situation with several possible accidents, total risk is the sum of the risks for each accident, provided that the outcomes are comparable:

$$\text{Total risk} = R_1 + R_2 + R_3 + \ldots + R_n$$

where $R_1, R_2, R_3 \ldots R_n$ represents the event risk. For example, if performing activity X has a probability of 0.01 of suffering an accident of type A, with a loss of 1000, and a probability of 0.000001 of suffering an accident of type B, with a loss of 2,000,000, the total risk is a loss of 12 based on a loss of 10 from an accident of type A plus 2 from an accident of type B.

One of the first major uses of this concept was the planning of the Delta Works flood protection program in the Netherlands in 1953 with the aid of mathematician David van Dantzig (Walman 2008). The kind of risk analysis pioneered for that project has become common today in fields like nuclear power, aerospace, and the chemical industry (Note 9).

---

## Fear as Intuitive Risk Assessment

People rely on their fear and hesitation to keep them out of unknown and possibly dangerous situations. De Becker (1997) argues that, "True fear is a gift for it is a survival signal that surfaces only in the presence of danger. Yet unaccountable fear has assumed a power over us that it holds over no other creature on Earth. It need not be this way. Risk could be said to be the way we collectively measure and share this *true fear*—a fusion of rational doubt, irrational fear, and a set of unquantifiable biases from our own experience."

For example, the field of behavioral finance focuses on human risk aversion, asymmetric regret, and other ways that human financial behavior varies from what analysts deem rational. Risk in this case is the degree of uncertainty associated with a return on an asset. Recognizing and respecting the irrational influences on human decision making may do much to reduce disasters caused by naive risk assessments that presume to rationality but in fact merely fuse many shared biases.

## Audit Risk

The audit risk model expresses the risk that an auditor will provide an inappropriate opinion of a commercial entity's financial statements. It can be analytically expressed as:

$$AR = IR \times CR \times DR$$

where AR is audit risk, IR is inherent risk, CR is control risk, and DR is detection risk.

## Other Considerations

Another consideration in risk management is that risks are future problems that can be treated, rather than current ones that must be immediately addressed.

## Risk versus Uncertainty

In his seminal work titled *Risk, Uncertainty, and Profit*, Knight (1921) established the distinction between risk and uncertainty: "Uncertainty must be taken in a sense radically distinct from the familiar notion of *risk*, from which it has never been properly separated. The term *risk* as loosely used in everyday speech and in economic discussion really covers two things which, functionally at least, in their causal relations to the phenomena of economic organization, are categorically different.... The essential fact is that *risk* means in some cases a quantity susceptible of measurement, while at other times it is something distinctly not of this character; and there are far-reaching and crucial differences in the bearings of the phenomenon, depending on which of the two is really present and operating.... It will appear that a measurable uncertainty, or *risk* proper, as we shall use the term, is so far different from an unmeasurable one that it is not in effect an uncertainty at all. We … accordingly restrict the term *uncertainty* to cases of the non-quantitative type." Thus, for Knight, uncertainty is immeasurable, not possible to calculate, while in his view risk is measurable.

Another distinction between risk and uncertainty was proposed by Hubbard (2007 p. 46, 2009 p. 39). Gertner (2003) and Lerner et al. (2000) suggested a similar distinction:

- Uncertainty: The lack of complete certainty, that is, the existence of more than one possibility. The true outcome, state, result, or value is not known.
- Measurement of uncertainty: A set of probabilities assigned to a set of possibilities, for example, "There is a 60% chance this market will double in 5 years."
- Risk: A state of uncertainty where some of the possibilities involve a loss, catastrophe, or other undesirable outcome.
- Measurement of risk: A set of possibilities each with quantified probabilities and losses, for example, "There is a 40% chance the proposed oil well will be dry with a loss of $12 million in exploratory drilling costs."

In this sense, Hubbard uses the terms so that one may have uncertainty without risk but not risk without uncertainty. We can be uncertain about the winner of a contest, but unless we have some a personal stake in the outcome, we face no risk. If we bet money on the outcome of the contest, then we incur a risk. In both cases, more than one outcome is possible. The measure of uncertainty refers only to the probabilities assigned to outcomes, while the measure of risk requires both probabilities for outcomes and losses quantified for outcomes.

## Risk Attitude, Appetite, and Tolerance

The *attitude, appetite,* and *tolerance* terms are often used similarly to describe an organization's or individual's attitude toward risk taking. *Risk averse, risk neutral,* and *risk seeking* terms may be used to describe a risk attitude. Risk tolerance looks at acceptable and/or unacceptable deviations from what is expected. Risk appetite looks at how much risk one is willing to accept. There can still be deviations within a risk appetite.

Gambling is a risk-increasing investment, wherein money on hand is risked for a possible large return, with the possibility of losing it all. Purchasing a lottery ticket is a very risky investment with a high chance of no return and a small chance of a very high return. In contrast, putting money in a bank at a defined rate of interest is a risk-averse action that yields a guaranteed small gain and precludes other investments with possibly higher gains. The possibility of getting no return on an investment is also known as the rate of ruin (Note 10).

## Risk as Vector Quantity

Hubbard argues that defining risk as the product of impact and probability presumes (probably incorrectly) that the decision makers are risk-neutral (Lerner and Keltner 2000). Only for a risk-neutral person is a certain monetary equivalent exactly equal to the probability of the loss times the amount of the loss. For example, a risk-neutral person would consider a 20% chance of winning $1 million exactly equal to $200,000 (or a 20% chance of losing $1 million to be exactly equal to losing $200,000). However, most decision makers are not actually risk-neutral and would not consider these equivalent choices. This concept led to the prospect theory and the cumulative prospect theory (Note 11).

Hubbard proposes instead that risk is a kind of vector quantity that does not collapse the probability and magnitude of a risk by presuming anything about the risk tolerance of the decision maker. Risks are simply described as a set or function of possible loss amounts, each associated with specific probabilities. How this array is collapsed into a single value cannot be determined until the risk tolerance of the decision maker is quantified.

Risks can be negative and positive, but people tend to focus on negative risks. This is because some activities can be dangerous, such as putting a life at risk. Risks concern people because people think risks will have negative effect on their futures.

## Disaster Prevention and Mitigation

We already discussed risk as the probability that a hazard will turn into a disaster. Vulnerabilities and hazards are not dangerous if taken separately. If they come together, they become a risk or, in other words, probabilities of disaster. Nevertheless, risks can be reduced or managed. If we are careful about how we treat the environment and understand our weaknesses and vulnerabilities to existing hazards, we can take measures to make sure that hazards do not turn into disasters.

Risk management goes beyond helping us prevent disasters. It also helps us to put into practice what is known as sustainable development. Development is sustainable when people can make a good living and be healthy and happy without damaging the environment or other people over the long term. For instance, you can make a living for a while by chopping down trees and selling the wood, but if you do not plant more trees than you cut down, you will eventually have no trees and no longer have the means to make a living, so this plan is not sustainable. Prevention and mitigation are actions we can take to make sure that a disaster doesn't happen or, if it does happen, it causes the least harm possible. We cannot stop most natural phenomena, but we can reduce the damage caused by an earthquake if we build stronger houses on solid ground.

What is prevention? Taking measures to prevent an event from turning into a disaster. Planting trees, for example, prevents erosion and landslides. It can also prevent drought. On the other hand, mitigation measures are activities that reduce vulnerability to certain hazards. For instance, certain building techniques ensure that houses, schools, and hospitals will not be knocked down by an earthquake or a hurricane. Prevention and mitigation begin with:

- Knowing which hazards and risks we are exposed to in our community.
- Getting together with our families and neighbors and making plans to reduce those hazards and risks and prevent them from harming us.
- Actually doing what we planned to do to reduce our vulnerability.
- Taking action, not just talking.

---

## Scenario Analysis

Scenario analysis is a process of analyzing possible future events by considering alternative possible outcomes (sometimes called *alternative worlds*). Thus, scenario analysis, which is a method of projection, does not try to show one exact picture of the future. Instead, it presents alternative future developments. Consequently, a scope of possible future outcomes is observable. Both the outcomes and the development paths leading to the outcomes are observable. In contrast to prognoses, scenario analysis does not involve extrapolation of the past or reliance on historical data; it does not expect past observations to remain valid in the future. Instead, it tries to consider possible developments and turning points that may be connected to the past. In short, several scenarios are demonstrated to show possible future outcomes.

The method is useful for generating optimistic, pessimistic, and most likely scenarios. Experience has shown that about three scenarios are most appropriate for further discussion and selection. More scenarios could make the analysis unclear (Aaker 2001 p. 108; Bea and Haas 2005 p. 279, 287).

Of course, the analysis is designed to allow improved decision making by allowing consideration of outcomes and their implications. Scenario analysis can also be used to illuminate "wild cards." For example, analysis of the possibility that the earth will be struck by a large celestial object (meteor) suggests that while the probability is low, the damage inflicted will be so high that the event is much more important (threatening) than the low probability in any one year would suggest. However, this possibility is usually disregarded by organizations using scenario analysis to develop a strategic plan since it has such overarching repercussions. (Special note: The meteor that hit Russia on February 16, 2013, may change this assessment and reasonable scenarios may develop for the future.)

In politics or geopolitics, scenario analysis involves modeling the possible alternative paths of a social or political environment and possibly diplomatic and war risks. For example, in the recent Iraq War, the Pentagon certainly had to model alternative possibilities that might arise in the war situation and had to position materiel and troops accordingly.

While there is value in weighting hypotheses and branching potential outcomes from them, reliance on scenario analysis without reporting some parameters of measurement accuracy (standard errors, confidence intervals of estimates, metadata, standardization and coding, weighting for nonresponse, error in reportage, sample design, case counts, etc.) is a poor second to traditional prediction. Especially for complex problems, factors and assumptions do not correlate in lockstep fashion. A specific sensitivity that is undefined may call an entire study into question.

It is faulty logic to think, when arbitrating results, that a better hypothesis will obviate the need for empiricism. In this respect, scenario analysis tries to defer statistical laws (e.g., Chebyshev's inequality law) because the decision rules occur outside a constrained setting. Outcomes are not permitted to just happen; rather, they are forced to conform to arbitrary hypotheses ex post, and therefore, there is no basis on which to place expected values. In truth, there are no ex ante expected values, only hypotheses; and one is left wondering about the roles of modeling and data decision. In short, comparisons of scenarios with outcomes are biased by not deferring to the data; this may be convenient, but it is indefensible.

We must emphasize here that *scenario analysis* is no substitute for complete and factual exposure of survey error in many studies (chemical, nuclear, automotive, aerospace, economic, and others). In traditional prediction, given the data used to model the problem, an analyst using a reasoned specification and technique can state within a certain percentage of statistical error the likelihood that a coefficient will fall within a certain numerical bound. This exactitude need not come at the expense of disaggregated statements of hypotheses. For example, R Software, specifically the "what-if" module (Stoll et al. 2006) has been developed for causal inference and to evaluate counterfactuals. These programs include fairly sophisticated treatments for determining model dependence, in order to state with precision how sensitive the results are to models not based on empirical evidence.

Finally, scenario analysis must not be confused with forecasting. Forecasting is a tool used to predict future events, but it uses calculations based on historical data or theoretical suppositions. Forecasting typically uses statistical data collected over time to project trends into the future. Scenarios, on the other hand, do not depend on statistics; they involve possible issues, concerns, problems, and failures. Generally, scenarios are based on "out of the box" thinking without any substantiation and utilize what-if and checklist analyses.

## Notes

### Note 1

ISO 31000 is intended to be a family of standards relating to risk management codified by the International Organization for Standardization (ISO). The purpose of ISO 31000:2009 is to provide principles and generic guidelines on risk management. ISO 31000 seeks to provide a universally recognized paradigm for practitioners and companies employing risk management processes to replace the myriad existing standards, methodologies, and paradigms that differ among industries, subject matters, and regions. Currently, the ISO 31000 family is expected to include:

- ISO 31000:2009—Principles and Guidelines on Implementation
- ISO/IEC 31010:2009—Risk Management and Risk Assessment Techniques
- ISO Guide 73:2009—Risk Management Vocabulary

### Note 2

The Open Group is a vendor and technology-neutral industry consortium, currently with over 400 member organizations. It was formed in 1996 when X/Open merged with the Open Software Foundation. Services provided include strategy, management, innovation and research, standards, certification, and test development.

### Note 3

*Cumulative incidence* or *incidence proportion* is a measure of frequency, as in epidemiology, where it is a measure of disease frequency over a period of time. If the period considered is an entire lifetime, the incidence proportion is called lifetime risk (Rychetnik 2004).

Cumulative incidence is defined as the probability that a particular event, such as the appearance of a particular disease, occurred before a given time (Dodge 2003). It is equivalent to the incidence, calculated using a time period during which all of the individuals in a population are considered to be at risk for the outcome. It is sometimes also referred to as the *incidence proportion*. Cumulative incidence is calculated by the number of new cases during a period divided by the number of subjects at risk in the population at the beginning of the study. It may also be calculated as the incidence rate multiplied by duration (Bouyer et al 2009):

$$CI(t) = 1 - e^{-R(t)*D}$$

## Note 4

*Risk aversion* is a concept in psychology, economics, and finance, based on the behavior of humans (especially consumers and investors) while exposed to uncertainty to attempt to reduce the uncertainty. Risk aversion is the reluctance of a person to accept a bargain with an uncertain payoff rather than another bargain with a more certain but possibly lower expected payoff. For example, a risk-averse investor might choose to put his or her money into a bank account with a low guaranteed interest rate rather than into a stock that may bring high returns but also involves a chance of losing value.

## Note 5

*Framing* in the social sciences refers to a set of concepts and theoretical perspectives on how individuals, groups, and societies organize, perceive, and communicate about reality. Framing is commonly used in risk analysis, media studies, sociology, psychology, and political science.

## Note 6

*Risk assessment* is a step in a risk management procedure. Risk assessment is the determination of qualitative or quantitative value of risk related to a concrete situation and a recognized threat (also called hazard). *Quantitative risk assessment* requires calculations of two components of risk (R): the magnitude of the potential loss *(L)* and the probability *(p)* that the loss will occur. In all types of engineering of complex systems, sophisticated risk assessments are often made in the areas of safety engineering and reliability engineering in relation to threats to life, environment, or machine functioning. The nuclear, aerospace, oil, rail, and military industries have a long history of dealing with risk assessment. Also, medical, hospital, and food industries control risks and perform risk assessments continually. Methods for assessment of risk may differ among industries and whether the assessment involves financial decisions or environmental, ecological, or public health risks.

*Risk management* is the identification, assessment, and prioritization of risks (defined in ISO 31000 as *the effect of uncertainty on objectives*, whether positive or negative) followed by a coordinated economical application of resources to minimize, monitor, and control the probabilities and/or impacts of unfortunate events (Hubbard 2009 p. 46) or maximize the realization of opportunities. Risks can arise from uncertainty in financial markets, project failures (at any phase in design, development, production, or sustainment life cycles), legal liabilities, credit risks, accidents, natural causes and disasters, deliberate attacks from adversaries, or events of uncertain or unpredictable root cause.

Several risk management standards have been developed by various organizations, including the Project Management Institute, the National Institute of Standards and Technology (NIST), actuarial societies, and the International Organization for Standardization (ISO) that created ISO/IEC Guide 73:2009 (2009) and ISO/DIS 31000 (2009). Methods, definitions, and goals vary widely according to whether the risk management method is in the context of project management, security, engineering, industrial processes, financial portfolios, actuarial assessments, or public health and safety. The strategies to manage risk typically include transferring the risk to another party, avoiding the risk, reducing its negative effect or probability, or even accepting some or all of the potential or actual consequences.

Certain aspects of many risk management standards have come under criticism for having no provision for measurable improvement even if the confidence in estimates and decisions seems to increase (ISO/DIS 31000: 2009).

### Note 7

The CCTA risk analysis and management method (CRAMM) was created in 1987 by the Central Computing and Telecommunications Agency (CCTA) of the United Kingdom government. CRAMM is currently on its fifth version. Each of its three stages is supported by objective questionnaires and guidelines. The first two stages identify and analyze the risks to a system. The third stage recommends how these risks should be managed. The three stages are as follows:

### Stage 1. Establishment of the objectives for security by:
- Defining the boundary for the study.
- Identifying and valuing the physical assets that form part of the system.
- Determining the value of the data by interviewing users about the potential business impacts that could arise from unavailability, destruction, disclosure, or modification.
- Identifying and valuing the software assets that form part of the system.

**Stage 2. Assessment of the risks** to the proposed system and the requirements for security by:

- Identifying and assessing the types and levels of threats that may affect the system.
- Assessing the extent of the system's vulnerabilities to the identified threats.
- Combining threat and vulnerability assessments with asset values to calculate measures of risks.

**Stage 3. Identification and selection of countermeasures** commensurate with the measures of risks calculated in Stage 2. CRAMM contains a very large library consisting of over 3000 detailed countermeasures organized into over 70 logical groupings.

### Note 8

The Methode Harmonisée d'Analyse de Risques (Method for Harmonized Analysis of Risks or MEHARI) was developed and distributed by CLUSIF (a group of French information security professionals). Since 1995, MEHARI has provided to information security personnel (ISO practitioners, risk managers, chief information officers, etc.) to enable them to evaluate and manage the risks attached to scenarios. MEHARI is derived from previous standards (ISO/IEC 13335) and steadily evolved to provide compliance to the newer ISO/IEC 27001-02 and ISO/IEC 27005 standards. MEHARI generally involves the analysis of the security stakes and a preliminary classification of the IS entities according to three basic security criteria (confidentiality, integrity, availability). The typical steps are:

- Involved parties list the dysfunctions that exert direct impacts on organization activity.
- Audits are conducted to identify potential information system (IS) vulnerabilities.
- The risk analysis is carried out.

MEHARI complies by design with ISO 13335 to manage risks. This method can thus take part in a stage of the information security management system (ISMS) model of ISO 27001 by:

- Identifying and evaluating the risks within the framework of a security policy (P).
- Providing precise information on the plans to be built (D) starting from reviews of the points of control of the vulnerabilities (C).
- Using a cyclic approach of piloting (A).

## Note 9

The Delta Works is a series of construction projects in the southwest of the Netherlands intended to protect a large area of land around the Rhine–Meuse–Scheldt delta from the sea. The works consist of dams, sluices, locks, levees, and storm surge barriers. The aim of the project was to shorten the Dutch coastline, thus reducing the number of dikes that had to be raised. Along with Zuiderzee Works, Delta Works has been declared one of the Seven Wonders of the Modern World by the American Society of Civil Engineers.

## Note 10

*Rate of ruin* is the probability that a trading stake will "go bust," based on a dollar equivalent standard deviation, a winner-to-loser ratio, and a dollar trading stake. The calculation utilizes the natural log, and the result is a percentage probability.

## Note 11

*Prospect theory* is a behavioral economic concept that describes decisions between alternatives that involve risk where the probabilities of outcomes are known. The theory says that people make decisions based on the potential value of losses and gains rather than the final outcome, and they evaluate the losses and gains using interesting heuristics. The model is descriptive: it tries to model real-life choices rather than optimal decisions. The paper titled "Prospect Theory: An Analysis of Decision under Risk" has been called a "seminal paper in behavioral economics" (Shafir and LeBoeuf 2002).

*Cumulative prospect theory (CPT)* is a model for descriptive decisions under risk that was introduced by Tversky and Kahneman (1992). It is a further development and variant of prospect theory. The difference between this version and the original prospect theory is that weighting is applied to the cumulative probability distribution function, as in rank-dependent expected utility theory but not applied to the probabilities of individual outcomes.

The main modification to prospect theory is that, as in rank-dependent expected utility theory, cumulative probabilities are transformed rather than individual probabilities. This leads to the overweighting of extreme events that occur with small probability rather than to overweighting of all small probability events. The modification helps avoid a violation of first-order stochastic dominance and makes the generalization to arbitrary outcome distributions easier. CPT is therefore on theoretical grounds an improvement over prospect theory.

# References

44 USC §3542(b)(1). http://www.law.cornell.edu/uscode/text/44/3542.

Aaker, D. (2001). *Strategic Market Management*. New York: John Wiley & Sons.

Ayyub, B. (2003). *Risk Analysis in Engineering and Economics*. New York: Chapman & Hall/CRC.

Bea, F. and J. Haas (2005). *Strategisches Management*. Stuttgart: Lucius & Lucius.

Bouyer, J., D. Hémon, S. Cordier et al. (2009). *Épidemiologie principes et méthodes quantitatives*. Paris: Lavoisier.

Carson, J., E. Elyasiani, and I. Mansur. (2008). Market risk, interest rate risk, and interdependencies in insurer stock returns: a system GARCH model. *Journal of Risk and Insurance*, 75, 873–891.

Chang, R., J. Schaperow, T. Ghosh, J. Barr, C. Tinkler and M. Stutzke. (January 2012). *State-of-the-Art Reactor Consequence Analyses (SOARCA) Report*. NRC publications in the NUREG series, NRC regulations, and *Title 10, Energy*, in the Code of *Federal Regulations*. Washington, DC: The Superintendent of Documents U.S. Government Printing Office Mail Stop SSOP.

Cortada, J. (2003). *The Digital Hand: How Computers Changed the Work of American Manufacturing, Transportation, and Retail Industries*. New York: Oxford University Press.

Cortada, J. (2005). *The Digital Hand, Volume II: How Computers Changed the Work of American Financial, Telecommunications, Media, and Entertainment Industries*. New York: Oxford University Press.

Cortada, J. (2007). *The Digital Hand, Volume III: How Computers Changed the Work of American Public Sector Industries*. New York: Oxford University Press, p. 496.

de Becker, G. (1997). *The Gift of Fear: Survival Signals That Protect Us from Violence*. Boston: Little, Brown.

Dodge, Y. (2003). *The Oxford Dictionary of Statistical Terms*. Cambridge: Oxford University Press.

Drake, R. (2004). Selective potentiation of proximal processes: neurobiological mechanisms for spread of activation. *Medical Science Monitor*, 10, 231–234.

Flyvbjerg, B. (2008). Curbing optimism bias and strategic misrepresentation in planning: reference class forecasting in practice. *European Planning Studies,* 16, 3–21.

Flyvbjerg, B. (2006). From Nobel Prize to project management: getting risks right. *Project Management Journal,* 37, 5–15.

Flyvbjerg, B. (2004). *Procedures for Dealing with Optimism Bias in Transport Planning: Guidance Document*. British Department for Transport. http://flyvbjerg.plan.aau.dk/0406DfT-UK%20OptBiasASPUBL.pdf

Flyvbjerg, B., B. Holm, and S. Buhl. (2005). How (in)accurate are demand forecasts in public works projects? *Journal of the American Planning Association,* 71, 131–146. http://flyvbjerg.plan.aau.dk/Traffic91PRINTJAPA.pdf

Flyvbjerg, B., B. Holm, and S. Buhl. (2002). Underestimating costs in public works projects: error or lie? *Journal of the American Planning Association,* 68, 279–295. http://flyvbjerg.plan.aau.dk/JAPAASPUBLISHED.pdf

Franklin, H. (2001). *The Science of Conjecture: Evidence and Probability before Pascal*. Baltimore: Johns Hopkins University Press.

Galasyuk, V. and V. Galasyuk. (2007). Consideration of economic risks in a valuation practice: journey from the Kingdom of Tradition to the Kingdom of Common Sense. Social Science Research Network. http://papers.ssrn.com/sol3/papers.cfm?abstract_id=1012812

Gertner, J. (2003). "What are we afraid of? *Money*, 32, 80.

Hubbard, D. (2007). *How to Measure Anything: Finding the Value of Intangibles in Business*. New York: John Wiley & Sons.

Hubbard, D. (2009). *The Failure of Risk Management: Why It's Broken and How to Fix It*. New York: John Wiley & Sons.

ISO/DIS 31000 (2009). *Risk Management: Principles and Guidelines on Implementation*. Geneva: International Organization for Standardization. http://www.iso.org/iso/iso_catalogue/catalogue_tc/catalogue_detail.htm?csnumber=43170

ISO/IEC 27005:2008. *Information Technology: Security Techniques: Information Security Risk Management*. Geneva: International Organization for Standardization.

ISO/IEC Guide 73:2009 *Risk Management Vocabulary*. Geneva: International Organization for Standardization. http://www.iso.org/iso/iso_catalogue/catalogue_ics/catalogue_detail_ics.htm?csnumber=44651.

ISO/IEC Guide 73:2002 *Risk Management Vocabulary*. Geneva: International Organization for Standardization.

ISO/IEC 27001. (2005). *Information technology: Security Techniques: Information Security Management Systems Requirements*. Geneva: International Organization for Standardization.

Jones, J. (2005). Introduction to Factor Analysis of Information Risk (FAIR). Risk Management Insight LLC. http://www.riskmanagementinsight.com/media/docs/FAIR_introduction.pdf

Knight, F. H. (1921) *Risk, Uncertainty, and Profit*. Chicago: Houghton, Mifflin.

Landsburg, S. (2003). Is your life worth $10 million? *Everyday Economics*. http://www.slate.com/id/2079475/ http://www.slate.com/articles/arts/everyday_economics/2003/03/is_your_life_worth_10_million.html

Lerner, J. and D. Keltner. (2000). Beyond valence: toward a model of emotion-specific influences on judgment and choice. *Cognition and Emotion*, 14, 473–493.

Luhmann, N. (1996). *Modern Society Shocked by Its Risks*. University of Hong Kong Department of Sociology Occasional Paper 17. Available via HKU Scholars HUB.

McElroy, T. and J. Seta. (2004). On the other hand, am I rational? hemisphere activation and the framing effect. *Brain and Cognition*, 55, 572–580.

Neill, M., J. Allen, N. Woodhead et al. (2008). A positive approach to risk requires person-centered thinking. *Tizard Learning Disability Review*. Person-Centered Risk Course Book. Stockport. HSA Press. http://www.thinklocalactpersonal.org.uk/_library/Resources/Personalisation/Personalisation_advice/A_Person_Centred_Approach_to_Risk.pdf

Nichols, R (2000). *Risk: Working Hypotheses*, 40th ed. New York: Houghton, Mifflin.

OHSAS 18001:2007. Risk is a combination of the likelihood of an occurrence of a hazardous event or exposure(s) and the severity of injury or ill health that can be caused by the event or exposure(s).

Rasmussen, N. C. (October 1975). "*Reactor safety study. An assessment of accident risks in U. S. commercial nuclear power plants.*" Executive Summary. *WASH-1400 (NUREG-75/014)*. Rockville, MD, USA: Federal Government of the United States, U.S. Nuclear Regulatory Commission. Washington, DC.

Rychetnik L, P. Hawe, E. Waters et al. (2004). A glossary for evidence based public health. *Journal of Epidemiology: Community Health*, 58, 538–545.

Sanderson, H. and J. Lewis. (2011). *A Practical Guide to Delivering Personalisation: Person-Centered Practice in Health and Social Care*. London: Jessica Kingsley Publishers.

Schatz, J., S. Craft, M. Koby et al. (2004). Asymmetries in visual-spatial processing following childhood stroke. *Neuropsychology*, 18, 340–352.

Schatz, J., S. Craft, M. Koby et al. (2004a). On the role of response conflicts and stimulus position for hemispheric differences in global/local processing: an ERP study. *Neuropsychologia*, 42, 1805–1813.

Shafir, E. and R. LeBoeuf. (2002). Rationality. *Annual Review of Psychology*, 53, 491–517.

Stoll, H., G. King, and L. Zeng. (2006). Whatif: software for evaluating counterfactuals. *Journal of Statistical Software*, 15, 1–19. http://www.jstatsoft.org/

Talbot, J. and M. Jakeman. (2009). *Security Risk Management Body of Knowledge*. New York: John Wiley & Sons.

Tversky, A. and D. Kahneman (1992). Advances in prospect theory: cumulative representation of uncertainty. *Journal of Risk and Uncertainty*, 5, 297–323.

Tversky, A. and D. Kahneman. (1981). The framing of decisions and the psychology of choice. *Science* 211 (4481), 453–458.

Wolman, D. (2008). Before the levees break: a plan to save the Netherlands. *Wired Magazine*. P. 3, Dec. 22.

## Selected Bibliography

Bernstein, P. (1998). *Against the Gods*. New York: John Wiley & Sons.

Clemens, P. and T. Pfitzer. (2006). Risk assessment and control. *Professional Safety*, 51, 41–44.

Department of the Army. (2006). *Composite Risk Management* FM 5-19 (FM 100-14). Washington.

Fahey, L. and R. Randall. (1997). *Learning from the Future: Competitive Foresight Scenarios*. New York: John Wiley & Sons.

Flyvbjerg, B., N. Bruzelius, and W. Rothengatter. (2003). *Megaprojects and Risk: An Anatomy of Ambition*. Cambridge: Cambridge University Press.

Franklin, J. (2001). *The Science of Conjecture: Evidence and Probability before Pascal*. Baltimore: Johns Hopkins University Press.

Gardner, D. (2008). *Risk: The Science and Politics of Fear*. New York: Random House.

Heldman, K. (2005). *Project Manager's Spotlight on Risk Management*. San Francisco: Jossey-Bass.

Hillson, D. (2007). *Practical Project Risk Management: The Atom Methodology*. Vienna, VA: Management Concepts.

Holton, G. (2004). Defining risk. *Financial Analysts Journal*, 60, 19–25. http://www.riskexpertise.com/papers/risk.pdf

Hopkin, P. (2012). *Fundamentals of Risk Management*, 2nd ed. London: Kogan-Page.

Kendrick, T. (2003). *Identifying and Managing Project Risk: Essential Tools for Failure-Proofing Your Project*. New York: American Management Association.

Linneman, R. and J. Kennell. (1977). Shirt-sleeve approach to long-range plans. *Harvard Business Review*, 55, 141.

Metzner-Szigeth, A. (2009). Contradictory approaches? on realism and constructivism in the social sciences research on risk, technology, and the environment. *Futures*, 41, 156–170.

Porteous, B. and P. Tapadar. (2005). *Economic Capital and Financial Risk Management for Financial Services Firms and Conglomerates*. New York: Palgrave Macmillan.

Proske, D. (2008). *Catalogue of Risks: Natural, Technical, Social, and Health Risks*. New York: Springer.

Rescher, N. (1983). *A Philosophical Introduction to the Theory of Risk Evaluation and Measurement*. Lanham, MD: University Press of America.

Schwartz, P. (1996). *The Art of the Long View: Paths to Strategic Insight for Yourself and Your Company*. New York: Random House.

# 2

## Approaches to Risk

Perhaps when one approaches risk in any situation, there is a profound need to differentiate disaster and mitigation. Risk is the probability that a hazard will turn into a disaster. Vulnerability and hazards taken separately are not dangerous, but if they come together, they become a risk or, in other words, the probability that a disaster will happen.

The last chapter indicates that probability allows risks to be reduced or managed. If we are careful about how we treat the environment and understand our weaknesses and vulnerabilities to existing hazards, we can take measures to ensure that hazards do not turn into disasters. Risk management helps us prevent disasters; it also helps us practice what is known as sustainable development. Development is sustainable when people can make good livings and be healthy and happy without damaging the environment or other people over the long term. As noted in Chapter 1, an individual can make a living for a while by chopping down trees and selling the wood, but the practice is not sustainable if he or she does not plant more trees than are cut down. The result will be no trees and no means to make a living.

We also touched on prevention and mitigation—actions we can take to ensure that a disaster does not happen or causes the least possible damage if it does. We cannot prevent most natural phenomena, but we can reduce the damage caused. For example, we can decrease earthquake damage by building stronger houses on solid ground. One way to address very complex risk issues is to evaluate them via *scenario analysis*, which is a process of analyzing possible future events by considering alternative possible outcomes (sometimes called alternative worlds). We also address this in the last chapter.

## Zero Mind-Set

When one deals with risk, the concerns about failures, accidents, and hazards require discussions by all parties concerned. Two fundamental concerns should be covered in order for discussions to be fruitful in the sense of eliminating or reducing failures, accidents, and hazards:

1. The zero accident mind-set
2. The principles of as low as reasonably practicable (ALARP) and so far as is reasonably practicable (SFAIRP)

Both issues are components of the risk planning (RP) activity that focuses on minimizing and/or eliminating all failures, accidents, and hazards. In essence, these points must be followed if the focus is on improving an organizations processes, operations, and equipment utilization.

To satisfy the philosophy of eliminating or reducing problems, a zero mind-set should be promoted and supported. Important elements of the mind-set are the understanding and implementation of the principles of ALARP based on trying to minimize all risks as far as practicable (and below the defined acceptable levels) after having assessed foreseen failure modes, consequences, and possible risk-reducing actions. ALARP generally is used to minimize both the probability of an undesired event and the consequences if it occurs. In practice, ALARP means that all personnel participating in preparation and execution of operations should actively seek to minimize risk as far as practicable through preventive operational planning and selecting safe solutions and robust designs. ALARP is considered a mind-set. Risk-reducing actions should be based on subjective cost–benefit assessments. Examples of such actions are the installation of critical low-cost components such as pad eyes and lifting gear, familiarization and hazard awareness training for personnel, and limiting numbers of personnel in potentially hazardous areas where wires are under tension.

While the basic philosophies of all safety and hazard programs are simple and easy to implement in any organization, the following cornerstones will make any hazard and safety program effective and improvement-driven:

- Every incident can be avoided.
- No job is worth getting hurt for.
- Every job will be done safely.
- Incidents can be managed.
- Safety is everyone's responsibility.
- Safety should be combined with best manufacturing practices.
- Safety standards, procedures, and practices must be developed.
- Training must ensure that everyone understands and meets safety requirements.
- Working safely is a condition of employment.

In addition to these basic requirements, the zero mind-set philosophy extends these characteristics to provide several benefits. The primary benefits are:

- Communication of safety standards to all employees
- Understanding and acceptance of responsibilities for implementing standards

- Documentation showing that standards and best management practices are met
- Internal management control
- Cost avoidance
- Improved quality
- Better productivity
- Team building
- Ability to identify unsafe behavior
- Failure to tolerate unsafe behavior
- Peer pressure that influences safe work practices
- Consistent planning and task execution

To maintain these benefits, it is imperative to have:

- Shared vision
- Cultural alignment
- Focus on incident control
- Upstream systems definitions
- Feedback
- Maintenance of safe attitude, awareness, action, and accountability (the four A's)
- Cultural change
- Commitments by all, especially management

## ALARP

In beginning a discussion about risk, invariably we start by explaining the concepts of *as low as reasonably practicable* (ALARP) and *so far as is reasonably practicable* (SFAIRP). Both terms have essentially the same meaning and the concept of *reasonably practicable* is at their cores. They both imply the need to weigh a risk against the trouble, time, and money needed to control it. Thus, ALARP describes the level to which we expect to see workplace risks controlled.

The use of the reasonably practicable principle allows us to set goals instead of prescriptive practices for the process owners. This flexibility is a great advantage, but it has drawbacks too. Deciding whether a risk is ALARP can be challenging because it requires all concerned to exercise judgment, often via a qualitative evaluation. Most cases can be decided by referring to existing good practices established by discussions with stakeholders to achieve a consensus about what constitutes ALARP. For highly hazardous, complex, or unique situations, we build on good practice using more formal decision-making techniques, including cost–benefit analysis on which to base our judgment.

In any health and safety system, the concept of reasonably practicable is of profound importance and serves as the center of the system. In Great Britain, the concept is a key element of the general duties imposed by the Health and Safety at Work Act of 1974 and part of many other health and safety regulations. It is imperative to ensure that the reader understands that all regulations must be based on what is reasonably practicable. By the same token, it may be impossible to follow the principle in some cases due to implementation directives of the European Union or other international body that uses similar but not identical concepts as ALARP.

The implementation of ALARP is fundamental to the work of an entire organization. Therefore, it is important that every employee at every level understands it:

- Policy makers and those engaged in program delivery must know about ALARP. When these parties develop proposals for any health, safety, and environmental actions to control health or safety risks, they must ensure as far as possible that the controls will reduce the risks to employees and others ALARP.

- Enforcers must understand ALARP because they decide whether the process owners reduced their risks ALARP and thus complied with the appropriate regulations and other pertinent standards.

- Technical specialists in health, safety, and environment must be knowledgeable about ALARP because they advise all parties concerned with health, safety, and the environment about whether control measures reduce risks ALARP, and they identify standards of risk control that are ALARP.

We briefly defined ALARP and SFAIRP but have not discussed them in terms of practicality. Let us look at *reasonable practicable* as it applies to both terms. SFAIRP is most often used in the Health and Safety at Work Act of 1974 and other regulations. ALARP is used more commonly by risk specialists and process owners. In most situations, ALARP is cited. However, in the health, safety, and environmental area, the consensus is that the terms are interchangeable with the exception that the correct phrase must be used in formal legal documents.

What is the correct legal phrase? According to the Britain's Court of Appeal definition in the case of *Edwards v. National Coal Board* [1949] 1 All ER 743, is:

> "Reasonably practicable" is a narrower term than "physically possible" … a computation must be made by the owner in which the quantum of risk is placed on one scale and the sacrifice involved in the measures necessary for averting the risk (whether in money, time, or trouble) is placed in the other, and that, if it be shown that there is a gross disproportion between them—the risk being insignificant in relation to the sacrifice—the defendants discharge the onus on them.

In essence, the court ensured that a risk reduced to ALARP met the standard of weighing the risk against the sacrifice needed to further reduce it. The decision is weighted in favor of health and safety because the presumption is that the process owners should implement the risk reduction measure. To avoid the need for this sacrifice, the process owners must be able to show that it would be grossly disproportionate to the benefits of risk reduction that would be achieved. Thus, the process is not one of balancing the costs and benefits of measures but rather of adopting measures except where they are ruled out because they involve grossly disproportionate sacrifices. Two extreme examples are (1) spending $1 million to prevent five staff members from suffering bruised knees is obviously grossly disproportionate; and (2) spending $1 million to prevent a major explosion capable of killing 150 people is obviously proportionate.

In reality, many decisions about risk and the controls to achieve ALARP are not so obvious because many factors come into play, such as ongoing costs set against remote chances of one-off events and the routine expenses and supervision time required to ensure that employees wear earplugs set against a chance of developing hearing losses in the future. ALARP requires judgment and no simple formula for computing it exists. The calculations may be very complicated. To facilitate the determination, the following checklist serves as a guideline:

- Assess compliance with the law and the use of good practice in each individual situation.
- Reduce risks by defining and understanding ALARP policy and guidance.
- Identify the principles and guidelines to assist the team in deciding whether process owners reduced risk ALARP.
- Identify health, safety, and environmental principles for a cost–benefit analysis (CBA) in support of ALARP decisions.
- Develop a cost–benefit analysis (CBA) checklist.
- Review ALARP in a cursory approach.

When dealing with ALARP, it is a common to confuse the meanings of *hazard* and *risk*. A hazard is an object, property of a substance, phenomenon, or activity that can cause adverse effects. For example:

- Water on a staircase is a hazard because an individual could slip, fall, and be injured.
- Loud noise is a hazard because it can cause hearing losses.
- Breathing asbestos dust is a hazard because it can cause cancer.

A risk, on the other hand, is the likelihood that a hazard will actually cause adverse effects and involves a measure of the effect. Risk is a two-part concept, and both parts are required for a practitioner to make sense of it. Likelihoods can be expressed as probabilities (one in a thousand), frequencies (1000 cases per year), or qualitatively (negligible, significant, etc.). The effect can be described in many ways. For example, we know that accidents happen in all areas of life. The following statistics are for the United Kingdom and the United States The sources are http://www.hse.gov.uk/education/statistics. htm and http://www.osha.gov/oshstats/commonstats.html, respectively.

### U.K. Statistics: Work Accidents Involving Young People between 1996 and 2001

- There were 54 deaths of young people (below age 18).
- There were 12,599 serious injuries involving broken limbs, amputations, and serious burns.
- There were 46,495 injuries leading to at least 3 days off work.

Safety issues produced even more alarming statistics:

- In 2005, 4 million people were hurt at work.
- In 2005, work injuries cost Britain £16 billion.
- Young people (aged 16 to 24) face the highest risks.
- New workers have the greatest risks of injuries.
- Average annual risk of injury is a direct consequence of an activity.

The following statistics are of interest on a per-year basis:

- Fairground accidents: 1 in 2,326,000 rides
- Road accidents: 1 in 1,432,000 km traveled
- Rail travel accidents: 1 in 1,533,000 passengers
- Burns and scalds in homes: 1 in 610

### U.S. Statistics (2011)

A total of 4,609 workers were killed on the job in 2011 (3.5 per 100,000 full-time equivalent workers)—almost 90 deaths per week or about 13 daily.

This represents a slight increase from the 4,551 work fatalities in 2009 and the second lowest annual total since the fatal injury census was first conducted in 1992. In the construction field, the following fatalities in 2011 are worth noting:

- Falls—251 of a total of 721 deaths (35%)
- Electrocutions—67 (9%)
- Struck by objects—73 (10%)
- Caught between objects—19 (3%)

Note that the 10 most frequently cited violations of OSHA standards in 2011 could have been minimized or even eliminated if proper analysis was performed. The standards in question are:

1. Scaffolding, general requirements, construction (29 CFR 1926.451)
2. Fall protection, construction (29 CFR 1926.501)
3. Hazard communication standard, general industry (29 CFR 1910.1200)
4. Respiratory protection, general industry (29 CFR 1910.134)
5. Control of hazardous energy (lock-out and tag-out), general industry (29 CFR 1910.147)
6. Electrical, wiring methods, components and equipment, general industry (29 CFR 1910.305)
7. Powered industrial trucks, general industry (29 CFR 1910.178)
8. Ladders, construction (29 CFR 1926.1053)
9. Electrical systems design, general requirements, general industry (29 CFR 1910.303)
10. Machine guarding, machines, general requirements, general industry (29 CFR 1910.212)

### How to Tell if a Risk Is ALARP

In the last section, we said that use of the reasonably practicable concept allows us to set goals for the process owners, rather than using prescriptive steps. This flexibility is a great advantage because it allows the process owners to choose the method that best suits them and supports innovation, but it has its drawbacks. Deciding whether a risk is ALARP can be challenging because it requires all involved parties to exercise judgment.

In most situations, deciding whether the risks are ALARP involves a comparison of existing or proposed control measures to the measures (good practices) we would normally expect to see in such circumstances. Good practice is generally defined as a set of standards for controlling risk that health, safety, and environmental authorities recognize as satisfying the law when applied to a particular case in an appropriate manner. The determination that an activity constitutes good practice must be based on a team consensus after a vigorous discussion involving all concerned parties (stakeholders).

Once good practice has been determined, much of the discussion with stakeholders about whether a risk is or will be ALARP is likely relate to the relevance of the practice and how appropriately it has been or will be implemented. When a good practice is deemed relevant, the responsible parties must follow it. If they want to pursue a different course, they must be able to demonstrate to the team's satisfaction that the measures they propose are at least as effective in controlling the risk. This is very important, and it must be determined via consensus and not through voting.

It is conceivable that deciding a good practice may be difficult for many reasons, including complexity of the issue or insufficient data for making a sound decision. In some cases, such as the introduction of new technology, the parties may have no relevant practice to compare. In that event, good practice should be followed as far as possible (usually with surrogate data), and the stakeholders should consider whether more steps can be taken to reduce the risk. If that is the case, the presumption is that process owners will implement these further measures, but this must be confirmed by going back to first principles to compare the risk with the sacrifice involved in further reducing it.

Often such first principle comparisons can be done qualitatively, i.e., by applying common sense and/or using professional judgment, theoretical knowledge, or experience. For example, if the costs are clearly very high and the reduction in risk is only marginal, it is likely that the situation is already ALARP and no more improvement is required. In other circumstances, improvements may be simple or economic to implement and the risk reduction significant. In this case, the existing situation is unlikely to be ALARP and the improvement is required. In many cases, decisions can be reached without further analysis.

Readers should understand that in some situations (highly hazardous industries or where new technology presents potentially serious consequences), decisions are less clear-cut and more detailed comparisons must be undertaken. The trouble is that risk and sacrifice are not usually measured in the same units and in essence involve comparing apples and flour. In these instances, a more formal cost–benefit analysis (CBA) may provide additional insight to help arrive at a judgment. However, a CBA alone does not constitute an ALARP case, cannot be used to argue against statutory duties, cannot justify intolerable risks, or justify poor engineering. A good solid CBA converts risk and sacrifice to a common set of units—money—to allow a comparison. It represents sacrifice as a cost and risk (insofar as it is reduced) as a benefit. Mathematically, the relationship is:

$$\frac{\text{Sacrifice}}{\text{Risk}} > 1 \times \text{DF, or } \frac{\text{Costs}}{\text{Benefits}} > 1 \times \text{DF}$$

where DF is the disproportion factor (a measure is not worth pursuing in comparison to the risk reduction achieved). DFs that may be considered gross vary 1 upward, depending on a number of factors, including the magnitude of the consequences and the frequency of realizing the consequences, i.e., the greater the risk, the greater the DF. Some general points for a CBA presented as part of an ALARP demonstration are:

- A CBA cannot be used to argue against the implementation of relevant good practice unless the alternative measures are demonstrated unequivocally to be at least as effective.
- The depth of analysis should be relevant and applicable for the identified purpose, i.e., more rigor is required where the risk is higher or the consequences (e.g., multiple fatalities) are great.
- A sensitivity analysis is usually required to support any conclusions suggesting that the costs are disproportionate to the benefits of implementing a measure.

As we can see from the formula above, the focus of a CBAs is to ensure that all the appropriate costs have been included and challenge the costs that seem out of the ordinary and excessive. Therefore, it would be proper to include the costs of installation, operation, training, additional maintenance, and the business losses that would follow a plant shutdown undertaken solely for the purpose of eliminating or minimizing a hazard. In fact, all claimed costs must be incurred by the process owner. Costs incurred by other parties such as members of the public should not be counted.

On the other hand, sacrifice implies non-recoverable cost. If a measure implies lost production, only the lost production during the delay can be counted. Conversely, if lost production is actually deferred production (if the life of the plant is based on operating time rather than calendar time), it should only take account of interest on the lost production plus allowance for operational costs during the implementation time and potential increase in operational costs at the end of life. For example, oil or gas remaining in a field while work is carried out on a platform should not be counted as lost production.

If the lost production costs strongly influence a decision not to implement, the process owner should show that phasing or scheduling the work to coincide with planned downtimes, for example, for maintenance, would not change the balance. The costs considered should be only those necessary and sufficient for the purpose of implementing the risk reduction measure (no "gold plating" or deluxe items).

Ongoing production losses as a result of the measure (slowed production or increased maintenance) can be counted. Any savings resulting from the measure such as reduced operational costs, avoidance of damage, and reinstatement costs should be offset against the above costs. These are not considered safety benefits but are counted as savings because they reduce the overall cost of implementing a measure. Finally, the costs claimed should relate only to the measure being implemented for safety. Translation into monetary costs is often uncertain and that is why all costs must be justified.

Now that we have looked at cost issues, we can consider the benefits. The focus is to ensure that all benefits of implementing a health and safety improvement measure are included and that the benefits associated with the measure are not underestimated. The benefits should include all reductions in risks to members of the public, workers, and the wider community. Benefits can be classified so that prevention is assured in typical areas such as (1) fatalities, (2) major and minor injuries, (3) ill health, and (4) environmental damage (control of major accident hazards or COMAH).

Benefits can also include avoidance of deployment of emergency services and avoidance of countermeasures such as evacuation and post-accident decontamination if appropriate. The cash valuations of preventing health and safety effects on people are presented in the Table 2.1. The value costs are estimates. It is very important to note that all benefits of reducing an injury type should be included. If a risk reduction measure is identified for one type of accident but reduces other risks such as health problems, all benefits should be counted. Because of this convolution, the responsible parties may need to treat reinstatement costs as benefits rather than offsetting them against costs. This would be the case if a plant being reinstated were safety-related plant, for example, one that treats hazardous wastes. This can

**TABLE 2.1**

Typical Cash Valuation for Cost–Benefit Analysis

| Type of Injury | Explanation | Value (Cost in $) |
|---|---|---|
| Fatality | Loss of life | 2,000,000 |
| *Injury* | | |
| Permanently incapacitating | Moderate to severe pain for 1 to 4 weeks; pain gradually reducing but may recur during some activities; some permanent restrictions to leisure and possibly work activities | 250,000 |
| Serious | Slight to moderate pain for 2 to 7 days followed by pain or discomfort for several weeks; some restrictions on work and leisure activities for several weeks to months; return to normal health after 3 to 4 months without permanent disability | 50,000 |
| Slight | Minor cuts and bruises; quick and complete recovery | 500 |
| *Illness* | | |
| Permanently incapacitating | Same as for injury | 250,000 |
| Other ill health | Absence exceeding 1 week; no permanent health consequences | 3,000 + 150 per day for absence |
| Minor | Absence up to 1 week; no permanent health consequences | 1,000 |

represent a bias *in favor* of safety because the gross disproportion factor is applied to all benefits prior to their comparison to the costs.

Now that we have looked both costs and benefits, it is time to analyze them. Obviously, a number of features within an analysis may influence on outcome. The following points should be considered when assessing the suitability of a CBA.

- Discounting of monetary values to translate future benefits and costs to present values is permitted.

- If future costs are significant, a process owner must consider discounting to see whether it may change the outcome of a finely balanced analysis. If a measure is deemed not reasonably practicable without discounting, the owner must show that the outcome would not differ if discounting was applied. Discounting of future costs, particularly if they are significant, may make a measure more favorable than if discounting was ignored because higher effective discount rates are applied to costs than to health and safety benefits.

- Future health and safety benefits should not be discounted at rates exceeding 3.25% (2012 figure based on http://www.bankrate.com/rates/interest-rates/prime-rate.aspx).

- Future costs and cost savings should be discounted at a rate not less than 3.25% (2012 figure based on http://www.bankrate.com/rates/interest-rates/prime-rate.aspx)

- Discount periods in excess of 50 years are problematic, and evidence that a measure is not indicated as a result of such an analysis feature should be viewed with caution.

- The analysis should be shown to be robust by appropriate sensitivity analyses in line with the precautionary approach. In particular, the results of any CBA associated with major accident hazards will be subject to uncertainty due to the need to estimate how severe and how often the accidents may occur. By their nature, such accidents are rare, but they produce severe consequences when they do happen.

- In some cases, the inputs to the CBA may have sensitivity ranges of factors of 3 or more. Unless the extreme value has been used in the analysis, an outcome where the gross DF was exceeded by less than this factor would not be a compelling indication that the improvement was not reasonably practicable. Responsible process owners should provide adequate justification that they used conservative inputs to the CBA or that the sensitivity range factors are appropriate.

- The analysis should justify an appropriate DF.

- In the event of a major accident, significant issues for responsible process owners to consider include (1) reputation, (2) share price, and (3) customer base and market share.

Although these issues are not ones that health, safety, and environment would require, a responsible process owner to consider they can often play a significant part of any decision about investing in new and safer technology. A strong warning to the reader is necessary. Just because the verbalization of the assumptions and uncertainties seems easy and practical, in practice the assumptions may be much more involved than discussed here. In such situations, the team should focus on aspects of risk analysis from several perspectives so that the outcome of a sound CBA may become one of several considerations involved in the judgment that a risk has been reduced ALARP. For example, in policy work and in operational work dealing with many hazards, you may also need to consider how the public feels about the risk. In such cases, a CBA should detail societal concerns in the areas of reducing risks and protecting people.

### Risk Leverage

Risk leverage analysis (RLA) is commonly included in preparation of a CBA. RLA measures the relative costs and benefits of performing various candidate risk resolution activities. The equation to calculate risk leverage is:

$$\frac{RE_{before} - RE_{after}}{Risk\ resolution\ cost}$$

where leverage is a rule for risk resolution that reduces risk by decreasing the risk exposure (RE). Risk resolution cost is the cost of implementing the risk action plan. The concept of leverage helps determine actions with the highest paybacks. Generally, major risk leverage exists in the early phases of all design and/or development projects. Of course, the intent is to identify as many risky items as early as possible to reduce rework costs and to minimize expensive fixes as the design and/or development moves into later phases.

The risk exposure preceding a specific activity is the probability of a cost overrun multiplied by the consequence of the overrun. At project completion, 100% of the cost overrun is the prior risk exposure. Assuming 75 cents on the dollar for a reasonable claim, only 25% of the cost overrun remains at risk after the damage claim activity. If the cost to prepare the entitlement (legal basis for the claim) and quantum (value of the claim) is $100,000 (3 experts for 6 months), the risk leverage would be 7.5 to 1 for a $1 million overrun, and 75 to 1 for a $10 million overrun (Hall 1998 p. 115). To calculate risk leverage, we use the following formula:

$$\frac{(100\ percent \times cost\ overrun) - (25\ percent \times cost\ overrun)}{Cost\ of\ claim}$$

## Failures, Accidents, and Hazards

We defined failures, accidents, and hazards. However, as they relate to the ALARP and SFAIRP concepts, we must understand the three basic causes for failures:

1. Human errors: acts of omission or commission by an operator, designer, contractor, or other person who creates a hazard that could result in a release of hazardous or flammable material or other event.

2. Equipment failures: a mechanical, structural, or operating failure results in the release of hazardous or flammable material or other event.

3. External events: items outside the unit reviewed affect the operation of the unit to the extent that the release of hazardous or flammable material or other event is possible. External events include upsets on adjacent units that impact the safe operation of the unit (or node) under study, loss of utilities, and exposures from weather and seismic activities.

Accidents are events that happen without warning and planning. Hazards are situations or events that threaten life, health, property, or environment. Failures, accidents, and hazards may all be defined as deviations from expected outcomes. The fact that they are deviations from goals or targets requires actions to resolve the deviations.

## ALARP Fallacies

Over the years, many myths have developed around ALARP. Many of these myths have spread over many applications in a variety of industries and organizations. The four most important ones are identified below (www. Hse.gov.uk).

**Ensuring that risks are reduced ALARP means that we have to raise standards continually**—It is part of health, safety, community, and environment philosophy that we seek continual improvements in health and safety standards. That philosophy is widely shared in several countries and produced excellent records whenever it has been applied. As a case in point, Britain has one of the best records for occupational health and safety in the world. However, for any country to achieve similar results, it must encourage improvements in a responsible way. Deciding whether an activity is safe enough (risk is reduced ALARP) is a separate exercise from seeking continual improvements in standards. Of course, as technology develops, new and better methods of risk control become available.

Process owners should review what is available from time to time and consider whether they need to implement new controls. That does not mean that the best risk controls available are always reasonably practicable. Only if the cost of implementing these new methods of control is not grossly disproportionate to the reduction in risk they achieve is their implementation reasonably practicable. For that reason, we accept that it may not be reasonably practicable to upgrade an older plant and equipment to modern standards. However, other measures such as partial upgrades or alternative measures may be required to reduce the risk ALARP.

The determination of what is ALARP will also be affected by changes in knowledge about the size or nature of the risk presented by a hazard. If sound evidence indicates that a hazard presents significantly greater risks than previously thought, of course we should press for stronger controls to handle the new situation. However, if the evidence shows the hazard presents significantly fewer risks than previously thought, we should accept a relaxation in control if the new arrangements ensure the risks are ALARP.

**If few employers have adopted a high standard of risk control, the standard is ALARP**—Some organizations implement standards of risk control that are more stringent than good practice for a number of reasons, such as meeting corporate social responsibility goals, striving to be the best, or reaching an agreement with staff to provide additional controls. It does not follow that these risk control standards are reasonably practicable simply just because a few organizations adopted them. Until such practices are evaluated and recognized by the properly recognized government authorities, an organization should not seek to enforce them at policy or operational level. It is also acceptable for a process owner to relax from a self-imposed higher standard to one accepted as ALARP, for example, simply meeting the requirements of a relevant approved code of practice (ACOP).

**Ensuring that risks are reduced ALARP means that we can insist on all possible risk controls**—ALARP does not mean that every measure that could possibly be taken (however theoretical) to reduce risk must be taken. Sometimes a risk can be controlled by more than one method. These controls can be considered barriers that prevent a risk from being realized. Companies are tempted to require more and more of these protective barriers to reduce risks as much as possible. Typical approaches may be controls such as limit switches, horns, light signals, and mistake proofing ("belt and braces" approaches). However, remember that ALARP means that a barrier can be required only if its introduction does not involve grossly disproportionate cost.

**Ensuring that risks are reduced ALARP means that there will be no accidents or ill health**—ALARP does not represent *zero risk*. We must expect a risk arising from a hazard to be realized on occasion and harm to occur even if the risk is ALARP. This is an uncomfortable thought, but it is inescapable. Of course we should strive to make sure that process owners reduce and maintain the risks ALARP, and we should never be complacent. However, we must accept the reality that risk from an activity can never be entirely eliminated unless the activity is stopped. This relates to the issue of risk tolerability and explains why risk assessments feed into contingency planning.

---

## Example

A simple method for coarse screening of measures is shown in Table 2.2. This puts the costs and benefits into a common format of dollars per year for the lifetime of a plant. Assume a distillery plant utilizing a process in which an explosion could lead to:

- 30 fatalities
- 50 permanent injuries
- 200 serious injuries
- 500 slight injuries

Further assume that the rate of occurrence of this explosion has been analyzed to be about $1 \times 10^{-6}$ per year or 1 in 1,000,000 annually. The plant has an estimated lifetime of 30 years. What we want to know is how much could the company reasonably spend to eliminate (reduce to zero) the risk from an explosion? If the risk of explosion were eliminated, the benefits can be assessed as shown in Table 2.2. The $3,390 represents the estimated benefit of eliminating a major accidental explosion at the plant on the basis of avoidance of casualties. The example did not include discounting or consider inflation. Also, for an injury to be recognized as *not* reasonably practicable, the cost must be grossly disproportionate to the benefits. This is taken into account by the disproportion factor (DF). In this case, the DF indicates the consequences of such explosions are great. A DF exceeding 10 is unlikely. In our example, it might be reasonably practicable to spend about $33,900

**TABLE 2.2**

Screening Measures

| Injuries | Number of Incidents | Value | Rate of Explosion | Years | Dollars |
|---|---|---|---|---|---|
| Fatalities | 30 | × 3,000,000 | × 1 × 10⁻⁶ | × 30 | 2,700 |
| Permanent injuries | 50 | × 250,000 | × 1 × 10⁻⁶ | × 30 | 375 |
| Serious injuries | 200 | × 50,000 | × 1 × 10⁻⁶ | × 30 | 300 |
| Slight injuries | 500 | × 1,000 | × 1 × 10⁻⁶ | × 30 | 15 |
| Total benefits | | | | | 3,390 |

($3,390 × 10) to eliminate the risk of an explosion. If a smaller DF is used, the responsible person must justify it.

This type of simple analysis can be used to eliminate or include some measures by costing various alternative methods of eliminating or reducing risks. Sometimes the numbers differ based on permanently incapacitating injuries and permanently incapacitating illnesses. This difference is justified because of the larger human cost attributed to injuries due to their short-term effects.

After we complete this preliminary analysis, the next step is to compare the sacrifice (cost) and the risk reduction (benefits). In a standard CBA, the usual rule is that a measure should be adopted only if the benefits outweigh the costs. However, in ALARP judgments, the rule is that a measure must be adopted unless the sacrifice is grossly disproportionate to the risk. Thus, the costs can outweigh benefits and the measure may still be reasonably practicable to introduce. How much costs can outweigh benefits before being judged grossly disproportionate depends on factors such as the magnitude of the risk; the larger the risk, the greater the disproportion between cost and risk.

### Differentiating Risks

Although risk is everywhere, those who deal with it must differentiate the levels for all tasks under investigation. Financial issues play a role in the specificity of a category, but also depend on the industry and magnitude of a project. The amounts cited here are only examples. Generally, risks are rated into three categories:

#### *Major Risks*

1. Employee fatalities or serious injuries during work on or off corporate premises
2. Contractor fatalities or serious injuries during work on or off corporate premises
3. Explosion, fire, or other acute incident resulting in significant damage to corporate or third-party property and/or third-party injuries
4. Third-party fatalities in incidents involving corporate vehicles
5. Major transportation accidents involving corporate products (e.g., vehicle rollover or product release)
6. Noteworthy product contamination (e.g., contaminated oxygen intended for medical use)
7. Property damage or business interruption likely to cost $1 million or more

8. Major incidents involving corporate staff, product, or property likely to receive significant media attention
9. Significant chemical spills that could pose threats to the environment

### Serious Risks

1. Lost time injuries or severe injuries without permanent disabilities
2. On-site material release contained with assistance
3. Off-site release with only minor detrimental effects
4. Statutory offense
5. Financial loss between $10,000 and $1 million
6. Media attention garnering local coverage

### Minor Risks

1. First aid or medical attention required
2. On-site material release immediately contained
3. Financial loss between $1,000 and $10,000

### Risk Priorities

Risk categories are very significant because they are components of a risk assessment. The three risk categories are generic, intended to explain risk in general. However, most corporations for simplicity use the following priorities (P's):

P1—Serious risk
P2—High risk
P3—Medium risk
P4—Low risk

When each risk has been assigned an appropriate reduction measure, the next step is to prioritize the actions. The system for prioritizing and setting target dates for hazards and risks varies based on type of organization, but a typical priority system may be:

Priority 1—Immediate risk control required, close-out within 1 month
Priority 2—Agreed plan within 1 month; complete within 6 months
Priority 3—Complete within 1 year
Priority 4—Action to be considered
Satisfactory—No action required

A priority rating should correspond to the level of risk; the greater the risk, the higher the priority. Other factors influencing a decision may require thorough reviews and documentation. Only executive management can approve extensions to P1 and P2 non-conformances. Some non-conformances are too large for corrective actions to be completed within the scheduled timelines. Extensions can be granted if a documented corrective action plan (CAP) showing amended timelines for completion is developed.

Priority may be defined in simpler terms for convenience when a project is not large or critical in nature. The alternative priority may be set as (1) high, (2), medium, (3) low, or (4) of no consequence. In addition to these qualitative measurements, it is very common in HAZOP analysis to use a risk assessment matrix (RAM) as shown in Table 2.3. Table 2.4 lists category definitions adopted from the Department of the Army.

**TABLE 2.3**

Risk Assessment Matrix: Hazard Probability

| Category | Definition | | | | |
|---|---|---|---|---|---|
| | **Frequent** | **Likely** | **Occasional** | **Seldom** | **Unlikely** |
| Catastrophic | E | E | H | H | M |
| Critical | E | H | H | M | L |
| Marginal | H | M | M | L | L |
| Negligible | M | L | L | L | L |

*Source:* Department of the Army (2006). *Composite Risk Management* FM 5-19 (FM 100-14). Washington, p. 8.

*Note:* E = extremely high. H = high. M = moderate. L = low.

**TABLE 2.4**

Categories of Risk Assessment Matrix

| Category | Description |
|---|---|
| Frequent | Frequent – Occurs very often, known to happen regularly |
| | Likely – Occurs several times, common occurrence |
| | Occasional – Occurs sporadically, not uncommon |
| | Seldom – Remotely possible, could occur at some time |
| | Unlikely – Probably will not occur, but not impossible |
| Catastrophic | Complete shutdown of project |
| | Death or permanent total disability |
| | Loss of major equipment |
| | Major property or facility damage |
| | Severe environmental damage |
| Critical | Severe downgrade of project status |
| | Permanent partial or temporary total disability |
| | Extensive major damage to equipment or systems |
| | Significant damage to property or environment |
| Marginal | Downgrade of project goals |
| | Minor damage to equipment, systems, property, or environment |
| | Lost days due to injury or illness |
| | Minor damage to property or environment |
| Negligible | Little or no adverse impact on project |
| | First aid or minor medical treatment |
| | Slight equipment or system damage, but fully functional or serviceable |
| | Little or no property or environmental damage |

# Reference

Hall, E. (1998). *Managing Risk.* New York: Addison Wesley.
http://www.hse.gov.uk/risk/theory/alarpglance.htm

# Selected Bibliography

Cagno, E., F. Caron, and M. Mancini. (2002). Risk analysis in plant commissioning: the multilevel HAZOP. *Reliability Engineering and System Safety, 77*, 309–323.
LaDuke, P. (2013). Bleeding money: how much does safety really cost? *Fabricating and Metal Working*, Feb., 18–19.
Linneman, R. and J. Kennell. (1977). Shirt-sleeve approach to long-range plans. *Harvard Business Review, 55*, 141.
Schwartz, P. (1996). *The Art of the Long View: Paths to Strategic Insight for Yourself and Your Company.* New York: Random House.
Topping, M. (2001). The role of occupational exposure limits in the control of workplace exposure to chemicals. *Occupational and Environmental Medicine, 58*, 138–144.

# 3

## *Types of Risk Methodologies*

There are many techniques to evaluate risk. In fact, the methodologies and specific tools vary as much as the organizations that use them. In this chapter, we will introduce some of the most common ones. Risk analysis methodologies fundamentally fall into three categories:

1. Qualitative methodologies
   a. Preliminary risk analysis
   b. Hazard and operability (HAZOP) studies
   c. Failure mode and effects analysis (FMEA); failure mode and effects criticality analysis (FMECA)
2. Tree-based techniques
   a. Fault tree analysis
   b. Event tree analysis
   c. Cause–consequence analysis
   d. Management oversight risk tree
   e. Safety management organization review
3. Techniques for dynamic systems
   a. Go method
   b. Digraph or fault graph
   c. Markov modeling
   d. Dynamic event logic analytical methodology
   e. Dynamic event tree analysis

## Qualitative Methodologies

### Preliminary Risk Analysis

Preliminary risk or hazard analysis is a qualitative technique involving a disciplined analysis of event sequences that could transform a potential hazard into an accident. The possible undesirable events are identified first and then analyzed separately. For each undesirable event or hazard, possible improvements or preventive measures are formulated.

The results provide a basis for determining which categories of hazards should be examined more closely and which analysis methods are most suitable. Such an analysis may also prove valuable in a working environment where activities lack safety measures that can be readily identified. With the aid of a frequency or consequence diagram, the identified hazards can then be ranked according to risk level, allowing measures to prevent accidents to be prioritized.

### Hazard and Operability (HAZOP) Studies

The HAZOP method is probably the most widely used analysis since two major accidents occurred in the petroleum and shipping industries (The *Exxon Valdez* oil spill in Alaska in 1989 and the British Petroleum spill in the Gulf of Mexico in 2010). Even those who are not familiar with the hazards analysis process may know the HAZOP term even if they are not really sure what it means. Chapter 5 is devoted exclusively to HAZOP studies. HAZOP can be defined as the application of a formal systematic critical examination of the process and engineering plans for new or existing facilities. The technique assesses the hazard potential arising from deviations in design specifications and the consequences faced by the operation or organization. The analysis involves the use of a set of guidewords and the development of scenarios intended to identify hazards or operational problems.

### Failure Mode and Effects Analysis (FMEA) and Failure Mode and Criticality Effects Analysis (FMCEA)

Failure mode and effects analysis (FMEA) was described at a very detailed level by Stamatis (2003). It is a bottom-up approach to identifying potential failures based on severity, occurrence, and detection. This chapter provides only a cursory summary.

The method was developed in the 1950s by reliability engineers to determine problems that could arise from malfunctions of military systems. FMEA analyzes each potential failure mode in a system, subsystem, or component to determine its effect on the system and classify it according to its severity. The purpose is to identify potential hazards associated with a design and/or process by investigating the failure modes for each component. FMEA is not effective for identifying hazards that involve the failures of multiple processes due to the complex interactions of the failures.

FMEA uses a similar approach to the what-if component of a HAZOP but has as its objective the identification of the effects of all the failure modes of each piece of equipment and instrumentation. As a result, FMEA identifies single failure modes that can play a significant part in an accident. It is not effective, however, for identifying combinations of equipment failures that lead to accidents. Human operators are not usually considered specifically in FMEA even though the effects of operational errors are usually included in equipment failure modes.

FMEA is similar to HAZOP technique but utilizes a different approach. HAZOP evaluates the impact of a deviation in operating conditions outside the design range (e.g., more flow or low temperature). FMEA uses a systematic approach to evaluate the impact of a single equipment failure or human error on a system or plant.

In FMEA, the reason or cause for the equipment failure is not specifically considered. In HAZOP, the causes for deviations must be assumed or confirmed by judgment and experience because HAZOP focuses on the causes. The FMEA methodology assumes that, if a failure can occur, it must be investigated and the consequences evaluated to verify whether the failure can be tolerated on safety grounds or whether the remaining serviceable equipment is capable of controlling a process safely.

To be effective, an FMEA needs a strong, well-led team with wide cumulative experience. The initial briefing by the leader and the contributions expected from each member are similar to the team requirements for a HAZOP. The results of the analysis are recorded as in a HAZOP. Table 3.1 is a typical initial record sheet. The recording should be in a standardized format for the whole plant in order to facilitate maintenance of records of the activities.

In carrying out the FMEA, the process flow diagrams and the process and instrumentation diagrams (P&IDs) are first studied to obtain a clear understanding of the plant operation. Where only part of a process is to be studied, it may be necessary to also include the failure modes of equipment immediately outside the analysis area and determine the consequences of those failures on the plant or process section being analyzed. Figure 3.1 depicts a typical flow for generating an FMEA. In performing a HAZOP analysis, it is common for a team to use a preliminary FMEA. Table 3.1 is a form for documenting this activity. Table 3.2 shows a second method of documenting the process.

**TABLE 3.1**

Initial FMEA Documentation

| Project: | Component: | | Page: | |
|---|---|---|---|---|
| Component description: | | | Date: | |
| | | | Drawing No. | |
| Number | Failure Mode | Detection Method | **Equipment Affected** | | Comments |
| | | | Identification | Effects | |
| | | | | | |
| | | | | | |
| | | | | | |
| | | | | | |

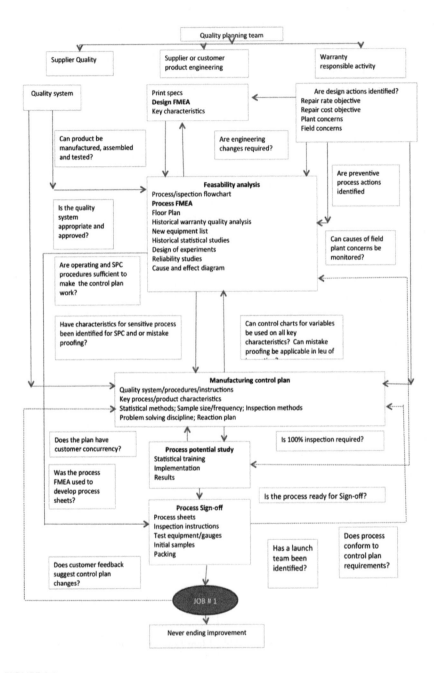

**FIGURE 3.1**
Typical flow for generating FMEA.

**TABLE 3.2**

Typical HAZOP/FMEA Worksheet

| FMEA Title: | | | | | FMEA No. | | | | |
|---|---|---|---|---|---|---|---|---|---|
| Project Title: System: Subsystem: Component: | | | | | Control No. Issue: | | | | |
| Prepared By: | | | | | Page ____ of ____ | | | | |
| Core Team: | | | | | | | | | |
| FMEA Type: Design Process | | | | | | | | | |
| Function or Requirement (Hazop Node or Item) | Potential Failure Mode (HAZOP Deviation) | Potential Effect (HAZOP Consequence) | S E V | Potential Cause) (HAZOP Cause) | O C C | Design or Process Control | D E T | R P N | Recommended Action | S E V | O C C | R P N |

| S = | O = | D = | R = |
|---|---|---|---|
| E = | C = | E = | P = |
| V = | C = | T = | N = |

The rationale for this activity is to make sure that the major items of concern are identified and the deviations are accounted for in the analysis of risk. When the FMEA is extended by a criticality analysis, the technique is then called failure mode and effects criticality analysis (FMECA). Table 3.3 is a typical combination worksheet. FMEA has gained wide acceptance by the automotive, aerospace, military, and many service industries. In fact, the technique has been adapted in other forms such as misuse mode and effects analysis and failure mode analysis.

### *Advantage*

FMEA involves a systematic review of a process. Its detailed methodology allows an item-by-item assessment of an operation.

### *Disadvantages*

FMEA may be time consuming and expensive. Complex processes will require investigation of many items, each involving examination of a complex series of failure modes.

FMEA is not effective for identifying hazards due to more than one failure. It is difficult to combine the effects of multiple failure modes of various items to identify combined hazards.

It is difficult to identify some items that may have been investigated to identify their various failure modes. Newer equipment may not be well documented, and some failure modes may be missed.

FMEA requires large amounts of data. The plant or operation needs to be well established before the technique can be performed, and the various failure rates for each item must be known.

### General Comments

The three techniques just discussed require the employment of hardware-familiar personnel only. FMEA tends to be more labor intensive because the failure of each component of a system must be considered. A point to note is that these qualitative techniques can be used in both the design and operational stages of a system.

In fact, all the techniques discussed have seen wide usage in the nuclear power and chemical processing plants. Furthermore, FMEA, one of the most documented methods, has been used in the automotive, aerospace, medical device, electronics, communication, and many more industries to improve the reliability of their processes and products. Figure 3.1 illustrates typical flow. Preliminary risk analysis has been applied to safety examination in industries and on offshore platforms. HAZOP is commonly used in the chemical industry to obtain detailed failure and effect data from studies of piping and instrumentation layouts and process and instrumentation layouts.

**TABLE 3.3**

FMEA and FMECA Worksheet

System
Subsystem
Component

| Activity No. | Equipment or Component Name | Function | Identification No. | Failure Mode | Intermediate Failure Effect | End Failure Effect | Failure Detection | Alternatives or Redundancies | Risk Value P × C = R (for FMECA only) | Comments |
|---|---|---|---|---|---|---|---|---|---|---|
| | | | | | | | | | | |
| | | | | | | | | | | |
| | | | | | | | | | | |
| | | | | | | | | | | |
| | | | | | | | | | | |

P = probability. C = consequence. R = risk value.

## Tree-Based Techniques

### Fault Tree Analysis (FTA)

A fault tree is a logical diagram that shows the relationship of system failure, i.e., a specific undesirable event in a system, and failures of the components of the system. It is a technique based on deductive logic. An undesirable event is first defined and the causal relationships of the failures leading to the event are identified. A fault tree can be used in qualitative or quantitative risk analysis. The difference is that the qualitative fault tree has a looser structure and does not require the same rigorous logic needed for a formal fault tree.

In essence, fault tree diagrams represent the logical relationship between a subsystem and component failures and how they combine to cause system failures. The top of a fault tree represents a system event of interest and is connected by logical gates to component failures known as basic events. After creating the diagram, failure and repair data are assigned to all system components. Analysis is performed to calculate the reliability and availability parameters of the system and identify critical components. Chapter 6 covers fault tree analysis.

### Event Tree Analysis (ETA)

ETA is discussed in detail in Chapter 7 and involves analysis of possible causes starting at system level and working down through subsystem, equipment, and component levels, identifying all possible causes. It determines what faults may be expected and how they occur. Assessment methods for quantifying the probability of an accident and the risk associated with plant operation based on a graphic description of accident sequences employ the FTA or ETA techniques.

ETA is also a logical method of analyzing how and why a disaster may occur. It is a great technique for working out the probability of a catastrophic event such as a nuclear power plant meltdown where the substantial cost of the analysis is obviously necessary. Both FTA and ETA can yield mathematical analyses of accident sequences and have been used to determine the reliabilities of electronic systems. FTA and ETA are also widely used in the nuclear industry but may not be suitable for general assessment of major hazards because they require substantial effort and cost.

ETA's tree structure provides a logical representation of the possible outcomes of a hazardous event. It also provides an inductive approach to reliability and risk assessment and is constructed using forward logic that allows the linkage of the ETA to a fault tree model by treating fault tree gate results as the sources of event tree probabilities.

## Cause–Consequence Analysis

Cause–consequence analysis (CCA) is a blend of FTA and ETA in that it combines cause analysis (described by fault trees) and consequence analysis (described by event trees) and hence uses deductive and inductive analyses. CCA is used to identify chains of events that can produce undesirable consequences. The probabilities of various events listed in a CCA diagram allow the calculation of the probabilities of the various consequences and ultimately establish the risk level of a system.

## Management Oversight Risk Tree (MORT) Analysis

MORT was developed in the early 1970s for the U.S. Energy Research and Development Administration as safety analysis method compatible with complex, goal-oriented management systems. A MORT diagram arranges safety program elements in an orderly and logical manner. The analysis is carried out via a fault tree whose top event consists of "damage, destruction, other costs, lost production, or reduced credibility of the enterprise in the eyes of society." The tree gives an overview of the causes of the top event arising from management oversights and omissions, assumed risks, or both.

The MORT tree compresses more than 1500 possible basic events to 100 generic events in the fields of accident prevention, administration, and management. MORT is used in the analysis or investigation of accidents and events and evaluation of safety programs. Its usefulness is revealed in the literature (Erickson 2005). Normal investigations revealed an average of 18 problems (and recommendations). Complementary investigations with MORT analysis revealed an additional 20 contributions per case.

## Safety Management Organization Review Technique (SMORT)

SMORT is a simplified modification of MORT developed in Scandinavia and structured by means of analysis levels with associated checklists (MORT is based on a comprehensive tree structure). Due to its structured analytical method, SMORT is classified as a tree-based method. SMORT analysis includes data collection based on the checklists and associated questions and evaluation of results. The information can be collected from interviews, studies of documents, and investigations. This technique can be used to perform detailed investigations of accidents and near misses and for performing safety audits and planning safety measures.

## General Comments

The tree-based methods are mainly used to determine conditions that lead to undesired events. In fact, ETA and FTA have been widely used to quantify the probabilities of accidents and other undesired events leading to the loss of life or economic losses in probabilistic risk assessment. However, their use is confined to static logic modeling of accident scenarios. Using ETA and FTA to analyze hardware failures and human errors does not allow explicit modeling of the conditions affecting human behavior. This affects the assessed level of dependency of events. Techniques such as human cognitive reliability to reconcile such deficiencies in FTA for modeling such responses have emerged and more will be developed in the future.

## Methodologies for Analysis of Dynamic Systems

### GO Method

The GO method is a success-oriented system analysis that uses 17 operators to aid in model construction. It was developed by Kaman Sciences Corporation in the 1960s to perform reliability analyses of electronics for the U.S. Department of Defense. The GO model can be constructed from engineering drawings by replacing system elements with one or more GO operators of three basic types: (1) independent, (2) dependent, and (3) logic.

Independent operators are used to model components requiring no input. Dependent operators require at least one input in order to generate an output. Logic operators combine the operators into the success logic of the system being modeled. The probability data for each independent and dependent operator allow the probability of successful operation to be calculated.

The GO method is used in practical application where the boundary conditions for a system to be modeled are well defined by schematics or other design documents. However, the failure modes are implicitly modeled, making the method unsuitable for detailed analysis of failure modes beyond the level of component events shown in the drawings. Furthermore, GO does not treat common cause failures nor provide structural information (minimum cut sets) for a system.

### Digraph or Fault Graph

The fault graph method or digraph matrix analysis uses the mathematics and language of graph theory such as *path set* (a set of models traveled on a path) and *reachability* (set of all possible paths between any two nodes). This method is similar to a GO chart but uses *and/or* gates instead. The connectivity matrix derived from adjacency matrix for the system shows whether a fault node will lead to the top event. These matrices are then subjected to computer analysis to yield *singletons* (single components that can cause system failure) or *doubletons* (pairs of components that can cause system failure).

This method allows cycles and feedback loops that make it attractive for dynamic systems.

## Markov Analysis (MA)

A Markov analysis is a state space technique that enables the modeling of failure probabilities and probabilities of state changes as well as failure and repair rates. It is one of the applied quantitatively methodologies using graphic models to represent possible states of a system under investigation, i.e., a component exhibits faultless operation, limited operation, or fails. However, one of the prerequisites of this deductive methodology is that the failure probabilities or probabilities of individual system states are known beforehand.

A typical assumption when using this analysis is that a system has no memory. Hence, the transitions between the individual states depend strictly on the current state and on time, but not on previous transitions between states. If a system exhibits different properties such as dependencies on further states, appropriate modeling actions must be taken that in some cases may significantly increase the complexity of the analysis. The dependency on time that may be considered in such analyses offers the possibility of considering failure and repair times. The time that elapses between the detection of a fault or failure and the repair of the element may be represented in a Markov model.

Markov analysis is based on the Markov process—a stochastic process with a finite number of states. Therefore, Markov analysis provides a method of analyzing the reliability and availability of subsystems representing components with strong interdependencies. Markov analysis is often used to model dependencies such as:

- Components in warm and/or cold standby
- Common maintenance personnel
- Limited on-site stocks of common spare parts

In practice, Markov models are used as sources of basic event data. In addition, they also may be analyzed independently of a fault tree analysis. In essence, Markov modeling is a classical modeling technique used for assessing the time-dependent behaviors of dynamic systems. In Markov chain processes, transitions between states are assumed to occur only at discrete points in time. In a discrete Markov process, transitions between states are allowed to occur at any time point. For process systems, the discrete method states can be defined in terms of ranges of process variables and component status.

This methodology incorporates time explicitly and can be extended to cover situations in which problem parameters are time independent. The state probabilities of a system *P(t)* in a continuous Markov system analysis are obtained by the solution of a coupled set of first-order, constant coefficient, differential equations:

$$dP/dt = MP(t)$$

where $M$ is the matrix of coefficients whose off-diagonal elements are the transition rates and whose diagonal elements are such that the matrix columns sum to zero. A typical application of Markov modeling is a hold-up tank problem. Pate-Cornell (1993) used the technique to study the fire propagation on a subsystem on board an off-shore platform.

### Dynamic Event Logic Analytical Methodology (DYLAM)

DYLAM provides an integrated framework to explicitly treat time, process variables, and system behaviors based on (1) component modeling, (2) system equation resolution algorithms, (3) setting of top conditions, and (4) event sequence generation and analysis. DYLAM is useful for describing dynamic incident scenarios and reliability assessment of systems whose missions are defined as values of process variables to be kept within certain limits in time. This technique can also be used to analyze system behavior and thus serve as a design tool for implementing protection and operator procedures.

It is important to note that a system-specific DYLAM simulator must be created to analyze each problem. Furthermore, input data such as the probability that a component will be in a certain state at transient initiation, the independence of such probabilities, the transition rates among different states, and the conditional probability matrices for dependencies among states and process variables must be provided to run the DYLAM package.

### Dynamic Event Tree Analysis Method (DETAM)

DETAM treats time-dependent evolution of plant hardware states, process variable values, and operator states over the course of a scenario. In general, a dynamic event tree is an event tree in which branching is allowed at different time points. The method is defined by five set characteristics: (a) branching set, (b) set of variables defining system state, (c) branching rules, (d) sequence expansion rule, and (e) quantification tools. The branching set refers to variables that determine the space of possible branches at any node on the tree. Branching rules, on the other hand, are used to determine a constant time step when branching should take place. The sequence expansion rules are used to limit the number of sequences.

This approach can be used to represent a wide variety of operator behaviors, model the consequences of operator actions, and serve as a framework for an analyst to employ a causal model for errors of commission. Thus, it allows the testing of emergency procedures and identifies where and how changes can be made to improve procedure effectiveness. A typical example of use is the analysis of an accident sequence for a steam generator tube rupture.

## General Comments

The techniques discussed above address the deficiencies found in FTA and ETA methodologies by analyzing dynamic scenarios. However, the techniques have limitations. The digraph and GO techniques model system behaviors and deal to a limited extent with changes in model structures over time. Markov modeling requires the explicit identification of possible system states and the transitions among them. However, it is difficult to envision a complete set of possible states prior to scenario development.

DYLAM and DETAM can solve the problem by defining implicit state transitions. The drawbacks to these implicit techniques are implementation-oriented. Computer resources are required for analysis of the large tree structures generated by DYLAM and DETAM. Another issue is that the implicit methodologies may require considerable analyst effort to gather data and construct models.

These 13 risk analysis techniques address some of the fundamental qualitative methodologies although they lack the ability to (1) account for the dependencies among events, and (2) effectively identify potential hazards and failures within a system. The tree-based techniques addressed this deficiency by considering the dependencies among events. The probability of occurrence of an undesired event can be quantified based on operational data. However, no one (to our knowledge) has yet attempted to quantify the undesired top event on a MORT tree.

Current research has utilized DYLAM and DETAM to study accident scenarios by treating time, process variables, system behaviors, and operator actions through an integrated framework. These techniques address the problem of less-than-adequate modeling of conditions affecting control system actions and operator behavior (e.g., behavior of plant process variables, decisions by operating personnel) when using FTA and ETA. However, two drawbacks for these techniques are the needs for extensive computer resources and large data collections. The development of more efficient algorithm and powerful computer processors allow these methods to be applied more widely.

## Traditional Methodologies

The methods described above are powerful ways of addressing risk and HAZOP issues. However, other less demanding approaches may also be used to identify these risks and the more typical and traditional ones are described below.

## What-If Method

The what-if method (the term is hyphenated and the question mark is omitted in the OSHA regulation) is the least structured hazard analysis technique and requires the least time. A what-if analysis is conducted by a team of experienced analysts, engineers, and operations experts whose knowledge and experience equips them to identify several scenarios in such a way that hazards may be eliminated or minimized. It is important to note that the team is relatively unstructured. The success of the analysis depends on the (1) knowledge, (2) thinking processes, (2) experiences, and (4) attitudes of the team members. This loosely defined approach allows the team members to be creative and to expand ideas for appropriate resolutions. In other words, the loose structure allows out-of-the-box thinking.

However, despite the lack of structure, all team members *must* prepare appropriately and thoroughly before they meet for discussion. This preparation permits and requires all members to participate actively and understand the issues at hand. Without this preparation, the team will end up discussing important issues rather than evaluating the findings of the team members in such a way that a resolution through consensus is agreed upon. Typical issues that can be discussed during a review include:

- Emergency shutdown systems
- Vents
- Flares
- Piping systems
- Electrical classification areas
- Truck, rail, ship, and barge movements
- Effluents and drains
- Noise
- Leaks
- Operating procedures
- Maintenance procedures
- Machinery, including cranes, hoists, and forklifts
- Public access and perimeter fencing
- Adjacent facilities
- Buried cables
- Overhead cables
- Special weather problems, including freezing, fog, winterization, rain, snow, ice, high tides, and high temperatures
- Toxicity of construction materials
- Demolition safety

A what-if analysis can be organized in one of two ways:

1. Divide the facility into nodes, similar to a HAZOP, except that the nodes are typically larger and more loosely defined. An example of node separation is shown in Figure I.3.
2. Organize the analysis by major items of equipment like an FMEA and then discuss the types of failure modes for each item.

Let us examine each approach separately. We begin with the node analysis and follow with guidelines for utilities, batch processes, operating procedures, and equipment layout.

Nodal analyses are usually organized around major sections of a process such as a distillation column or a launching system. Team members ask what-if questions such as: What if there is high pressure? What if the operator forgets to do this? What if there is an external fire in this area? Using this approach, many of the individuals on the team will probably instinctively follow the HAZOP guideword approach. Consequently, a what-if analysis of this type may take the form of a faster-than-normal HAZOP. However, the scribe of the team will not need to take notes on every deviation guideword. Only meaningful discussions should be recorded. What-if discussions tend to jump from node to node more than normal in a HAZOP analysis, thus placing greater pressure on the leader and scribe to achieve results and arrive at relevant conclusions. Some what-if questions for as nodal analysis are:

- What if the system is bypassed?
- What if the flow stops?
- What if contamination occurs?
- What if a power failure occurs?
- What if there is corrosion or erosion?
- What if there is an external impact?
- What if the operator fails to pay attention?
- What if the operator skips a step?
- What if an instrument error occurs?
- What if an interlock is bypassed?

In the second approach to a what-if analysis, the discussions are organized around equipment types and their functions, for example:

- Pressure vessels
- Pumps
- Compressors
- Distillation columns

- Absorbers
- Storage tanks
- Vents
- Flares
- Piping systems

The concept is to use the what-if questions to deal with issues such as leaks and over-pressure related to specific equipment types. In the case of utility systems, the analysis of steam headers and instrument air systems can be difficult because the locations of nodal boundaries are not always clear. A discussion that starts in one area can roam far and extend almost across an entire facility. It is very important to recognize that utility systems involve large numbers of process interfaces, any of which may leak.

Sometimes a leak will be from a utility into a process; or the leak may be from a process to a utility. The source of a problem can be difficult to detect in either situation. To optimize detection, one way of analyzing a system is for the team leader and scribe to note potential interface problems as they are discussed during the process analysis. These notes can then be discussed by the group when the utilities are analyzed.

In the case of batch processes, hazard analysis methodologies were developed initially for large continuous operations such as petrochemical plants and refineries. However, as we learned more about the methodology, we applied the HAZOP principles to smaller organizations, especially those that utilize batch processes in pharmaceutical production and food processing. In some cases, we apply the batch process principles to continuous operations with internal batch operations such as truck loading and unloading.

The operation of a batch process is dynamic and time is a variable. Therefore, the analysis of a batch process is more complex than analyzing a steady-state operation. One way of handling this additional complexity is to systematically work through the operating procedures using a what-if approach in which deviation guidewords prompt questions. For example, if an instruction is to add 200 liters of water to V-200, the team might ask:

| Question | Guideword(s) |
| --- | --- |
| What if the vessel is over-filled? | High level |
| What if a liquid is not water? | Contamination |
| What if fewer than 200 liters are available? | Low flow |
| What if V-200 is over-pressured? | High pressure |
| What if the water is added too soon? | High flow |
| What if the water is added too late? | Low flow |
| What if the step is omitted altogether? | Low flow |

After the discussion for this step is complete, the team can analyze the next step of the operating procedure (OP). Other issues for discussion are

(1) whether a step is done early; (2) whether a step is done late; and (3) whether a step was omitted. OPs represent another way of looking at hazards and evaluating them appropriately. OPs are sometimes called standard operating procedures (SOPs). They identify specific procedures needed to achieve a particular task. It is common in hazard analysis to evaluate procedures for completeness and accuracy. If the procedures are not complete or accurate a hazard may result, especially if the procedure is not reflected in the totality of the process system. A what-if approach is an effective method of conducting such an analysis. The team works through each step of the procedure by asking a series of what-if questions:

1. What if the instruction is missed, overlooked, or ignored?
2. What if two instructions are followed in the wrong order?
3. What if a step is performed out of sequence (too early)?
4. What if a step is performed out of sequence (too late)?
5. What if a step is done too slowly?
6. What if a step is done too quickly?
7. What if an instruction is carried out partially (e.g., a valve is only partly closed)?
8. Does the operator have the information needed to perform this step? Can all relevant gauges be read?
9. Can this step be performed at night?

Finally, when determining risks, layout considerations must be evaluated. Equipment layout is an important issue for both productivity and safety. Matters to consider include:

- Ease of escape in the event of a fire or other serious event
- Noise zones
- Vehicle movements
- Accessibility for emergency vehicles
- Objects dropped from cranes and other lifting equipment

## Checklist

The checklist method uses a set of prepared questions to stimulate thinking, often in the form of a what-if discussion. The questions are developed by experts experienced with hazards analysis and the design, operation, and maintenance of process facilities. Checklists are never all-inclusive because no one can predict all options and hazards. As a result of this limitation, no hazard analysis method can make the claim that it is foolproof method and can foresee all hazards. Although this is a major limitation of the method,

**TABLE 3.4**

Topics for Generating Checklist Questions

| Checklist Question Topics | Items of Concern | Checklist Question Topics | Items of Concern |
|---|---|---|---|
| Equipment | Pumps<br>Compressors<br>Pressure vessels<br>Storage tanks<br>Piping<br>Valves | Control loops | Emergency loops |
| Utilities | Steam (various pressure levels)<br>Cooling water<br>Refrigerated water<br>Process or service water<br>Instrument air<br>Service air<br>Boiler feed water<br>Nitrogen<br>Other utility gases<br>Fuel gas<br>Natural gas<br>Electrical power<br>Firefighting equipment<br>External fire equipment<br>Runaway reaction prevention | Emergency systems | Fire water |
| Pressure relief | Relief valves<br>Rupture disks<br>Flares and flare headers | Human factors | Operating procedures<br>Training |
| Control loops | Emergency loops | Chemicals | Toxicity<br>Flammability<br>Corrosivity |
| Emergency systems | Fire water<br>Firefighting equipment<br>External fire equipment<br>Runaway reaction prevention | Instruments and controls | Local instruments<br>Board-mounted instruments<br>Distributed control system (DCS) |

a checklist should be broad and complete in covering questions that upon review will instill confidence that nothing obvious has been overlooked.

Although checklists represent a limited approach, the reality is that they are used in all types of hazards analysis. For example, checklists dealing with equipment failures are used in FMEAs. Examples of topics for generating checklist questions are listed in Table 3.4. Appendix A contains checklist questions dealing with safety, facility, and health issues.

A checklist generally has two sections as shown in Table 3.5. This example (based on www.stb07.com) identifies issues and concerns for chemical

**TABLE 3.5**

Sample Chemical Storage Checklist

| Company | | | | | |
|---|---|---|---|---|---|
| Facility | | | | | |
| Location | | | | | |
| Persons interviewed | Name: | | Title: | | Date: |
| | | | | | |
| | | | | | |
| Documents reviewed | Title: | | | | Date: |
| | | | | | |
| | | | | | |
| Notes | | | | | |

| # | Question | Yes, No, or Not Applicable | Notes |
|---|---|---|---|
| 1 | Are chemicals separated according to the following categories? | | |
| | Solvents, including flammable and combustible liquids, and halogenated hydrocarbons | | |
| | Inorganic mineral acids (nitric, sulfuric, hydrochloric, and acetic acids); bases (sodium hydroxide, ammonium hydroxide) | | |
| | Oxidizers | | |
| | Poisons | | |
| | Explosives or unstable reactives | | |
| 2 | Are caps and lids on all chemical containers tightly closed to prevent evaporation of contents? | | |
| 3 | Are material safety data sheets (MSDSs) provided for all chemicals at the facility? | | |
| 4 | Are hazardous chemicals purchased in the smallest quantities possible? | | |
| 5 | Are the MSDSs readily accessible? | | |
| 6 | Is a Hazardous Materials Team in place? | | |
| 7 | Are all chemicals properly logged in on receipt? | | |
| 8 | Is a list of chemicals on hand maintained at all times? | | |

*Continued*

**TABLE 3.5 (Continued)**

Sample Chemical Storage Checklist

| # | Question | Yes, No, or Not Applicable | Notes |
|---|----------|---------------------------|-------|
| 9 | Are all chemical containers properly labeled? | | |
| 10 | Is the safety diamond system used? | | |
| 11 | How are chemicals brought into the facility checked? | | |
| 12 | Are flammable or toxic chemicals stored near accommodation or office areas? | | |
| 13 | Are chemical drums and totes lifted over areas where people are present? | | |
| 14 | Are chemicals stored on stable flooring? | | |
| 15 | Are chemical storage areas properly vented? | | |
| 16 | Are chemicals ever stored in a refrigerator used for food? | | |
| 17 | Are storage shelves large enough? | | |
| 18 | Are storage shelves secure? | | |
| 19 | Do storage shelves have proper lips? | | |
| 20 | Are island shelf assemblies avoided? | | |
| 21 | Are procedures in place for responding to chemical spills in chemical storage areas? | | |
| 22 | Is the storage area constructed of flammable materials? | | |
| 23 | Does the storage area have an effective fire, smoke, and gas warning system? | | |
| 24 | Does the storage area have an effective fire control system? | | |
| 25 | Are incompatible chemicals stored in the same area? | | |

storage. The top section provides information as to how the checklist will be used. The company, facility, and location are identified. If some of the information for the checklist answers comes from discussions and interviews with personnel at the site, their names are noted. The titles of all documents reviewed are also entered in the top section. The bottom section lists the questions. The answer choices are yes, no, or not applicable. Discussions and background data may be entered in the Notes column.

## What-If and Checklist Combination

This combination method is the third of the hazards analysis techniques listed in the OSHA standard. It is basically a combination of the two methods that have just been introduced. The hazards analysis team works through a checklist. However, instead of merely answering yes or no to the questions, the team leader generates a relatively unstructured what-if discussion around each question.

## Indexing

Comparative risk levels can be evaluated using indexing methods. Each design is scored on a variety of factors contributing to overall risk. For example, a design that uses highly toxic chemicals will score negative points, whereas a facility located away from populated areas receives positive points. Credit is also provided for the use of control and mitigation measures. Three commonly used indexing methods are:

1. The Dow Fire and Explosion Index described in *Dow's Fire & Explosion Index Hazard Classification Guide*, 7th ed. (1994). American Institute of Chemical Engineers.
2. The Dow Chemical Exposure Index described in *Dow's Chemical Exposure Index Guide*. (1998). American Institute of Chemical Engineers.
3. The Pipeline Risk Management Index. In Muhlbauer, W. *Pipeline Risk Management Manual, 3rd ed.* (2003) Maryland Heights, MO. Gulf Professional Publishing.

## Interface Hazards Analysis

Most hazards analyses review subsets of larger systems. For example, a refinery hazards analysis team may perform a hazards analysis on only a catalytic cracking unit; a pipeline company may analyze only marine loading operations; or an offshore team may analyze only one platform of a larger complex. However, these subsets and subsystems are components or larger systems and thus can transfer hazards to or from other units across the interfaces.

Sutton (2010) reports on one large oil production facility that conducted both onshore and offshore operations. An operator performing a routine pigging operation on a line from an offshore platform to the onshore gas processing plant inadvertently misaligned the valves around the pig trap and caused a high-pressure surge to flow back along the line from the offshore segment. The mishap had no significant effect on the onshore operations, but the pressure surge caused the offshore platform to shut down, triggering a

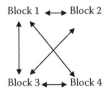

**FIGURE 3.2**
Interconnectivity. (*Source:* Ian Sutton. http://www.stb07.com/process-safety-management/
process-hazards-analysis.html. With permission.)

chain reaction that caused many other offshore platforms in the complex to
shut down in sequence. In the end, many millions of dollars of production
were lost, and the company was lucky no safety or environmental incident
occurred. Because management and the technical staff had not conducted an
*interface* hazards analysis, they did not understand the interactions of various operating units.

Another example of interface operations concerns truck operations. Many
process facilities use trucks from third-party companies to deliver chemicals
and export products and waste streams. It is generally a good idea to invite a
representative of a trucking company to a pertinent process hazards analysis.
That way both parties can be assured that the chances of mishaps are small.
The process facility, for example, can evaluate the procedures to ensure that
delivered chemicals are what they should be; the trucking company representative can check for the possibility of reverse flow of process chemicals
onto his company's trucks.

Sutton also reports that no established methodology exists for analyzing
system connectivity—for conducting what is in effect an IHA. Such a system
can be viewed as a collection of black boxes; each black box represents an
operating unit, each of which has been thoroughly analyzed individually.

Furthermore, Sutton (2010) has shown (Figure 3.2) a system consisting of
four operating units, each of which can be connected to all the others in some
manner, except that there is no link between Block 2 and Block 4. The arrows
that point two ways indicate that connectivity problems can flow in either
direction. For a system containing N blocks, the total number of connections
is $2 \times 3 \times (N - 1)$. The 2 represents a two-way connection. The 3 represents
the three types of connections (process fluids, instrument signals, and people noted below). In the case of Figure 3.2, the total of potential interfaces is
$2 \times 3 \times 3 = 36$ (30 if the missing connection between 2 and 4 is considered).

An interface hazards analysis (IHA) normally covers three areas:

1. Process fluids (e.g., incorrect analyses, reverse flows, incorrect
   compositions)
2. Instrument signals
3. People

One way of conducting the analysis is with the what-if approach. A hazards analysis team can use a flowchart of the overall process to ask what-if questions such as:

What if the flow in this pipeline suddenly stops?

Can the Unit A operators shut down any equipment on Unit B (instrumentation issue)?

What does Unit B do if Unit A has a fire (human communication and response issue)?

At each interface, the analyst will ask questions such as:

How do we know?

What is the consequence?

Are the safeguards adequate?

What is the effect of an upset on other units?

It is important not to draw too sharp a line between the methods. Indeed, the more experience a person gains in conducting and leading hazard analyses, the more the techniques seem to merge with one another. No single method is inherently better than any of the others. They all have their benefits and specific uses. A very good discussion of interfaces appears in Sutton (2010), Chapters 3 and 4 (and at www.stb07.com/process-safety-management/process-hazards-analysis.html).

## References

Erickson, C. II. (2005). Hazard Analysis Techniques for System Safety. New York: John Wiley & Sons, Inc.

http://www.stb07.com/process-safety-management/process-hazards-analysis.html

Paté-Cornell, E. (1993). Risk analysis and risk management for offshore platforms: lessons learned from the Piper Alpha accident. *Journal of Offshore Mechanics and Arctic Engineering*, 115, 179–190.

Stamatis, D. (2003). *Failure Mode and Effect Analysis (FMEA): From Theory to Execution.* Milwaukee, WI: Quality Press.

Sutton, I. (2010). *Process Risk and Reliability Management.* New York: Elsevier.

# 4

## Preliminary Hazard Analysis (PHA)

The material in this chapter should help a design team perform a preliminary hazard analysis (PHA). PHA is a design tool that helps engineers identify and deal with hazards in the initial stages of design. Performing a PHA allows engineers and management to better recognize and correct the hazards associated with designs for plants, units, and/or equipment. For an overview of a PHA, see Figure 4.1. Specifically, a PHA is a qualitative analysis performed in the earliest stages of design primarily to:

1. Identify all potential hazards and accidental events that may lead to an accident
2. Rank the identified accidental events according to their severity
3. Identify required hazard controls and follow-up actions
4. Formulate appropriate measures to deal with hazards

These four expectations may be also identified and rearranged in an order of mitigation goals:

1. Eliminate hazard
2. Control hazard with design methods
3. Incorporate safety devices to control hazard
4. Provide warning devices if hazard materializes
5. Provide procedures and training for operators

Other approaches may also be used to evaluate hazards. The most common ones are (1) rapid risk ranking and (2) hazard identification (HAZID). Although a PHA is conducted very early in the design, the subsequent benefits warrant the effort. The three major benefits are:

1. Product safety is ensured.
2. Modifications are less expensive and easier to implement in early design stages.
3. Design time is decreased because "surprises" are minimized.

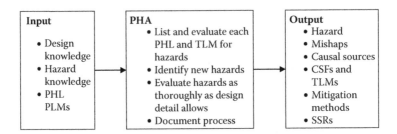

**FIGURE 4.1**
PHA overview.

The two most common limitations of the technique are:

1. Hazards must be foreseen by the designers.
2. The effects of interactions among hazards are not easily recognized.

PHA generally can be used as an initial risk study in an early stage of a project such as constructing a new plant. Accidents are mainly caused by releases of energy. A PHA indicates where energy may be released and which accidental events may occur and roughly estimates the severity of each accidental event. PHA results are used to (1) compare main concepts, (2) focus on important risk issues, and (3) yield inputs for more detailed risk analyses.

PHA can serve as the first step of a detailed risk analysis of an existing or planned system or concept. The purpose in that case is to identify accidental events that should be subject to more detailed risk analyses. A third use of PHA is to perform a complete risk analysis of a rather simple system. Whether a PHA will be a sufficient analysis depends on the complexity of the system and the objectives of the analysis. An effective PHA should consider:

- Hazardous components
- Safety-related interfaces among various system elements, including software
- Environmental constraints, including operating environments
- Operating conditions, testing, maintenance, built-in-tests, diagnostics, and emergency procedures
- Facilities, real property, installed equipment, support equipment, and training
- Safety-related equipment, safeguards, and possible alternate approaches
- Malfunctions to systems, subsystems, or software

**TABLE 4.1**

Typical Severity and Probability Classifications

| Hazard Severity Classification | Description | Accident Probability Classification | Description |
|---|---|---|---|
| Catastrophic | Causes multiple injuries, fatalities, or loss of facility | Probable | Likely to occur immediately or within a short time |
| Critical | May cause severe injury, severe occupational illness, or major property damage | Reasonable probable | Probably will occur |
| Marginal | May cause minor injury, minor occupational illness resulting in lost workdays, or minor property damage | Remote | Possibly may occur |
| Negligible | May not affect safety or health of personnel, but violates a safety or health standard | Extremely remote | Unlikely to occur |

The process of conducting a PHA is very simple and utilizes five steps:

1. Identify known hazards.
2. Determine their causes.
3. Determine their effects.
4. Determine the probability that an accident will be caused by a hazard.
5. Establish initial design and procedural requirements to eliminate or control hazards.

As important as these steps are, they are meaningless unless they are associated with severity and probability data for each event. Table 4.1 depicts common associations of severity and probability.

The simplicity and qualitative approach of PHA makes it difficult to determine what kind of hazards should be evaluated under its rubric. The following list of common sources may be helpful in making that determination:

- Sources and propagation paths of stored energy in electrical, chemical, or mechanical form
- Mechanical moving parts
- Material or system incompatibilities
- Nuclear radiation
- Electromagnetic radiation (including infrared, ultraviolet, laser, radar, and radio frequencies)

- Collisions and subsequent problems of survival and escape
- Fire and explosion
- Escapes of toxic and corrosive liquids and gases from containers or generated of such chemicals by other incidents
- Deterioration in long-term storage
- Noise, including subsonic and supersonic vibrations
- Biological hazards, including bacterial growth in such places as fuel tanks
- Human error in operating, handling, or moving near equipment
- Software error that can cause accidents

Obviously, the list is not exhaustive, but it should help readers determine areas that present potential hazards. Table 4.2 lists additional categories that may aid in developing a checklist and finding additional sources of hazards. Table 4.3 is a PHA worksheet based on Hammer (1989 p. 555). Table 4.4 presents a preliminary hazard matrix to help identify potential failures (Vincoli 1993 p. 68).

Information may be presented in many ways. In addition to PHA, a team may use the Table 4.5 format that is convenient and easy to use, especially as a means for documenting brainstorming activities. After a PHA is complete, a team should consider at least four post-PHA design activities:

1. Establish procedures to ensure that hazard elimination or control measures are effectively incorporated into the design.
2. Prepare a hazard report for each hazard (see Table 4.6).
3. Verify that the design eliminates or adequately controls the hazard.
4. Sign off on the hazard report (see Table 4.6).

Periodically, reviewing a job hazard analysis ensures that it remains current and continues to help reduce workplace accidents and injuries. Even if a job has not changed, it is possible that a team will identify hazards not identified by the initial analysis during a post-PHA review.

It is very important to review the job hazard analysis if an illness or injury occurs on a specific job. Based on the circumstances, one may determine that it necessary to change a procedure to prevent similar incidents in the future. If an employee's failure to follow proper job procedures results in a close call, discuss the situation with all employees who perform the job and remind them of proper procedures. Whenever a job analysis reveals a need for a revision, it is important to train all employees affected by the changes in methods, procedures, or protective measures.

**TABLE 4.2**

Typical Source for PHA Checklist

**System Operation:**

| Evaluator: | | Date: | |
|---|---|---|---|
| **Category** | **Item** | **Category** | **Item** |
| Electrical | Shock | Mechanical | Sharp edges or points |
| | Burn | | Rotating equipment |
| | Overheating | | Reciprocating equipment |
| | Ignition of combustibles | | Pinch points |
| | Inadvertent activation | | Lifting weights |
| | Power outage | | Stability and topping |
| | Distribution feedback | | potential |
| | Unsafe failure to operate | | Ejected parts and |
| | Explosion, electrical | | fragments |
| | (electrostatic) | | Crushing surfaces |
| | Explosion, electrical (arcing) | | |
| Pneumatic and | Overpressurization | Acceleration, | Inadvertent motion |
| hydraulic | Pipe, vessel, or duct rupture | deceleration, | Loose object translation |
| pressures | Implosion | gravity | Impacts |
| | Mislocated relief device | | Falling objects |
| | Dynamic pressure loading | | Fragments or missiles |
| | Improperly set relief | | Sloshing liquids |
| | pressure | | Slips and trips |
| | Back flow | | Falls |
| | Cross flow | | |
| | Hydraulic ram | | |
| | Inadvertent release | | |
| | Miscalibrated relief device | | |
| | Blown objects | | |
| | Pipe or hose whip | | |
| | Blast | | |
| Temperature | Heat source or sink | Ionizing | Alpha |
| extremes | Hot or cold surface burns | radiation | Beta |
| | Pressure evaluation | | Neutron |
| | Confined gas or liquid | | Gamma |
| | Elevated flammability | | X-Ray |
| | Elevated volatility | | |
| | Elevated reactivity | | |
| | Freezing | | |
| | Humidity or moisture | | |
| | Reduced reliability | | |
| | Altered structural properties | | |
| | (e.g., embrittlement) | | |
| Fire and | Fuel | Nonionizing | Laser |
| flammability | Ignition source | radiation | Infrared |
| factors | Oxidizer | | Microwave |
| present | Propellant | | Ultraviolet |

*Continued*

**TABLE 4.2 (Continued)**

Typical Source for PHA Checklist

| Category | Item | Category | Item |
|---|---|---|---|
| Explosives and effects | Mass fire<br>Blast overpressure<br>Thrown fragments<br>Seismic ground wave<br>Meteorological reinforcement | Explosive initiators | Heat<br>Friction<br>Impact or shock<br>Vibration<br>Electrostatic discharge<br>Chemical contamination<br>Lightning<br>Stray welding sparks |
| Explosive sensitizers | Heat or cold<br>Vibration<br>Impact or shock<br>Low humidity<br>Chemical contamination | Explosive conditions | Explosive propellant<br>Explosive gas<br>Explosive liquid<br>Explosive vapor<br>Explosive dust |
| Materials arising from leaks and spills | Liquids or cryogens<br>Gases or vapors<br>Irritating dusts<br>Radiation sources<br>Flammable<br>Toxic<br>Reactive<br>Corrosive<br>Slippery<br>Odorous<br>Pathogenic<br>Asphyxiating<br>Flooding<br>Run-off<br>Vapor propagation | Chemical and water contamination | System cross connection<br>Leaks and spills<br>Vessel, pipe, or conduit rupture<br>Backflow or siphon effect |
| Physiological | Temperature extremes<br>Nuisance dust or odor<br>Barometric pressure extreme<br>Fatigue<br>Lifted weights<br>Noise<br>Vibration (Raynaud's syndrome)<br>Mutagens<br>Asphyxiants<br>Allergens<br>Pathogens<br>Radiation<br>Cryogens<br>Carcinogens<br>Teratogens<br>Toxins<br>Irritants | Human factors | Operator error<br>Inadvertent operation<br>Failure to operate<br>Early or late operation<br>Out-of-sequence operation<br>Right operation, wrong control<br>Operated too long<br>Operated too briefly |

**TABLE 4.2 (Continued)**

Typical Source for PHA Checklist

| Category | Item | Category | Item |
|---|---|---|---|
| Ergonomic | Fatigue | Control systems | Power outage |
| | Inaccessibility | | Electromagnetic or |
| | Inadequate or no kill | | electrostatic interference |
| | switches | | Moisture |
| | Glare | | Sneak circuit |
| | Inadequate control or | | Sneak software |
| | readout | | Lightning strike |
| | Differentiation | | Grounding failure |
| | Inappropriate control or | | Inadvertent activation |
| | readout | | |
| | Location | | |
| | Faulty or inadequate control | | |
| | readout | | |
| | Labeling | | |
| | Faulty workstation design | | |
| | Inadequate or improper | | |
| | illumination | | |
| Unannunciated | Electricity | Common | Utility outage |
| Utility Outage | Steam | Causes | Moisture and humidity |
| | Heating or cooling | | Temperature extremes |
| | Ventilation | | Seismic disturbance or |
| | Air conditioning | | impact |
| | Compressed air or gas | | Vibration |
| | Lubrication drains and | | Flooding |
| | slumps | | Dust and dirt |
| | Fuel | | Faulty calibration |
| | Exhaust | | Fire |
| | | | Single-operator coupling |
| | | | Location |
| | | | Radiation |
| | | | Wear |
| | | | Maintenance error |
| | | | Vermin, varmints, mud |
| | | | daubers |

*Continued*

**TABLE 4.2 (Continued)**

Typical Source for PHA Checklist

| Category | Item | Category | Item |
|---|---|---|---|
| Contingencies (emergency responses by systems or operators to unusual events) | Hard shutdown or failure<br>Freeze<br>Fire<br>Windstorm<br>Hailstorm<br>Utility outage<br>Flood<br>Earthquake<br>Snow and ice load | Mission phasing | Transport<br>Delivery<br>Installation<br>Calibration<br>Checkout<br>Shakedown<br>Activation<br>Standard start<br>Emergency start<br>Normal operation<br>Load change<br>Coupling and uncoupling<br>Stressed operation<br>Standard shutdown<br>Emergency shutdown<br>Diagnosis and trouble shooting<br>Maintenance<br>Pressure gauge<br>Heating coil<br>Plug<br>Thermostat |

*Note:* No hazards checklist should be considered complete. Every list should be enlarged as experience and specific applications require. This list intentionally contains redundant entries because redundancy may lead to a different point of view in an open discussion.

## Example PHA: Home Electric Pressure Cooker

Most pressure cookers include safety devices of the following types:

1. A safety valve to relieve pressure before it reaches dangerous levels
2. A thermostat to open the circuit through the heating coil when the temperature exceeds 250°C
3. A pressure gauge divided into green and red sections; the red section indicates danger

**TABLE 4.3**

PHA Worksheet

| System or Subsystem Function | | | | | | Analyst: | | | | |
|---|---|---|---|---|---|---|---|---|---|---|
| Preliminary Hazard Analysis | | | | | | Date: | | | | |
| Number | Hazard | Causes | Effects | Mode | IMRI | Recommended Action | FMRI | Comments | Responsible Individual | Status |
| | | | | | | | | | | |
| | | | | | | | | | | |
| | | | | | | | | | | |

IMRI = No mitigation techniques or safety requirements available.
FMRI = Mitigation techniques and/or safety requirements available.

**TABLE 4.4**

Preliminary Hazard Matrix

| Hazard Group | Potential Failure Area | | | | | |
|---|---|---|---|---|---|---|
| | Structural | Procedural | Electrical | Mechanical | Pressure | Leakage/Spill |
| Collision/ mechanical damage | | | | | | |
| Loss of habitable atmosphere | | | | | | |
| Corrosion | | | | | | |
| Contamination | | | | | | |
| Electric shock | | | | | | |
| Fire | | | | | | |
| Pathological impact | | | | | | |
| Psychological impact | | | | | | |
| Temperature extreme | | | | | | |
| Radiation | | | | | | |
| Explosion | | | | | | |
| System Operator: | | | | | | |
| Evaluator: | | | | Date: | | |

**TABLE 4.5**

Typical PHA Brainstorming Record

| Project Component | Incident Type | Scenario | Proposed Treatment Measure | Likelihood | Consequence | Risk |
|---|---|---|---|---|---|---|
| | | | | | | |
| | | | | | | |
| | | | | | | |
| | | | | | | |
| | | | | | | |
| | | | | | | |
| | | | | | | |

**TABLE 4.6**

Typical PHA Report

| Title: | Report no.: |
|---|---|
| Equipment or system: | Report date: |
| Close-out date: | Authorized signature: |
| Authorizing individual: | |
| Description of hazard and accident that may result: | |
| Events and conditions that may contribute to hazard or accident: | |
| Possible means to eliminate or control hazard or accident effects: | |
| Estimated probability of accident occurrence: | |
| Current condition with control:<br>• Frequent<br>• Reasonably probable<br>• Occasional<br>• Remote<br>• Extremely improbable | |
| Explain choice: | |
| Means of verifying adequacy of control or applicable safety requirements: | |
| Organization or person to take action: | |
| Status of action already or to be or taken: | |

A typical PHA will consider any hazards associated with a design during the earliest stages of the design process. Two concerns are:

1. Appropriate measures for dealing with hazards should be incorporated into a design.
2. Subsequent hazard analyses should then be performed as the design progresses in order to identify new hazards and assess the ability of the design to minimize their harmful effects. FMEA, FMECA, and FTA may also be used to assess and minimize design hazards.

The intent is to design a safe product. A PHA will help ensure that:

1. A final product is safe by helping designers identify and deal with hazards.
2. Modifications are made in the early design stages because they are less costly and easier to implement than modifications made later.
3. Designers can anticipate hazards, thereby reducing the number of surprises that arise during the design process; in many cases, an effective PHA may expedite the design process.

An effective team assigned to develop a PHA must:

1. Identify known hazards through a preliminary hazards matrix that will divide hazards into generic groups and associate potential failures with the hazard groups.
2. Generate a hazards checklist to identify specific hazards based on the experiences of the team, theoretical knowledge of the process, and/or:
   a. Equipment descriptions
   b. Incident and accident report data
   c. Past operational history of similar tasks
   d. Review of other historical records

Note that no hazards checklist should be considered complete because subsequent hazards may appear during the design process. At this point, the causes, severity of the effects that may affect personnel, equipment, facilities and/or operations, and appropriate probabilities should be discussed. Forms such as Tables 4.1 through 4.5 may be used. A single hazard may involve numerous possible causes. The PHA at this stage should attempt to identify all possible causes. The causes of hazardous conditions will often become more apparent as the details of the design are better defined. Furthermore, the failure of one part of a system may cause failures of other parts. The PHA should estimate the overall effects of a hazard or failure.

### Severity and Probability

The severities of the effects of hazards generally fall into four categories:

- Catastrophic: may cause multiple injuries, fatalities, or loss of a facility
- Critical: may cause severe injury, severe occupational illness, or major property damage
- Marginal: may cause minor injury, minor occupational illness resulting in lost workdays, or minor property damage
- Negligible: probably will not affect the safety or health of personnel, but violates a safety or health standard

Generally, estimates of the probability of an accident in the early design stages are very subjective. The usual probabilities are defined as:

- Probable: likely to occur immediately or within a short time
- Reasonably probable: probably will occur
- Remote: may occur
- Extremely remote: unlikely

The next step is establishing initial (or revised if applicable) design and procedural requirements to eliminate or control the hazards. Post-PH activities involve establishing procedures to ensure that hazard elimination or control measures are effectively incorporated into a design.

A hazard report (see Table 4.6) may be created for each new hazard identified during the design process. The report may be used to track a hazard through the design process to ensure that appropriate measures are incorporated into the design to eliminate or adequately control the hazard. Of course, the ability of the design to eliminate or control every identified hazard must be verified by test results. The report may be signed off only after a design effectively eliminates or adequately controls the hazard.

### PHA Limitations

A PHA will only be as effective as the design team's ability to recognize hazards. If a hazard is not recognized, the PHA will be no help in minimizing it. A PHA does not effectively account for interactions among hazards.

### Preventive and Corrective Measures

When a PHA is complete recommendations are made to correct and to prevent failures and minimize the associated risks. Some hazards, recommendations, and preventive actions surrounding the design and use of a pressure cooker are:

- Shock: faulty wire insulation creates a circuit to ground through a user when the user touches the cord. Mild shock and even electrocution can result, depending on the resistance to current flow through the user's body. The overall resistance depends on factors such as the resistance from the user's, whether the user's fingers were wet, and the condition of the device insulation.
- Insulation should be resistant to deterioration.
- The device should include a grounded (three-prong) plug and be plugged only into an outlet equipped with a ground-fault circuit interrupter.
- Fire sparks are generated near flammable material when current passes from the cord to another object at a point where the insulation is faulty. Significant damage to system and surroundings can result.
- Extremely remote probability: a fault in insulation generates sparks that reach a flammable material in close proximity to the cord resulting in fire or shock.
- Flammable materials should be kept away from device.

- Burns occur when a person touches a hot pressure cooker surface, hot contents, or steam from a safety valve. The duration of contact with a hot surface or material determines whether burn severity is first or second degree.
- Hot pads or mitts should be used if the pressure cooker must be touched.
- Keep the pressure cooker out of the reach of children.
- A cover on the safety valve will spread the steam flow so that it is not concentrated enough to burn the skin.
- Explosion can occur when the thermostat and safety valve fail and no one notices that the pressure gauge indicates danger. An explosion can cause severe injuries or fatalities, destruction of the device, and damage to surroundings. The possibility of explosion is remote if high-quality thermostats and safety valves are used. Redundancies such as two safety valves also help prevent accidents.

## References

Hammer, W. (1989). *Occupational Safety Management and Engineering*, 4th ed. Englewood Cliffs, NJ: Prentice Hall.

Vincoli, J. (1993). *Basic Guide to System Safety*. New York: Van Nostrand Reinhold.

## Selected Bibliography

Hoxie, W. (2003). Preconstruction risk assessments. *Professional Safety*, 48, 50–53.

Smith, K. and D. Whittle. (2001). Six steps to effectively update and revalidate PHAs. *Chemical Engineering Progress*, 97, 70–77.

# 5

## HAZOP Analysis

### Overview

Hazard and operability (HAZOP) analysis is a structured and systematic technique for system examination and risk management. It is often used to identify potential hazards in a system and operability problems likely to lead to non-conforming products. HAZOP is based on a theory that assumes risk events are caused by deviations from design or operating intentions. Identification of such deviations is facilitated by using guidewords as a systematic list of deviation perspectives. This approach is a unique feature of the HAZOP methodology that helps stimulate the imaginations of team members when exploring potential deviations. As a risk assessment tool, HAZOP is often described as:

- A brainstorming technique
- A qualitative risk assessment tool
- An inductive risk assessment tool; a bottom-up risk identification approach in which success relies on the ability of subject matter experts (SMEs) to predict deviations based on past experiences and their general expertise

The International Conference on Harmonisation (ICH) Q9 Guideline titled *Quality Risk Management* endorses the use of HAZOP as one of many alternative tools for managing pharmaceutical quality risk and is also used in marine operations and environmental studies. In addition, to its utility in quality risk management, HAZOP is also commonly used to assess industrial and environmental health and safety applications. Additional details on the HAZOP methodology may be found in IEC International Standard 61882, *Hazard and Operability Studies (HAZOP) Application Guide*.

Just as in successful application of any tool or methodology, including risk management model, the tools must be appropriate and applicable to the problem at hand. This chapter discusses the principles of HAZOP in the context of the accepted quality risk management process, consisting of risk assessment, risk control, risk review, and communication and is intended to complement (not replace or repeat) the guidance available from IEC International Standard 61882. This discussion will focus on the basic principles of HAZOP and follow with a detailed analysis of the principles and the execution of a typical analysis.

## Definitions

HAZOP methodology requires understanding of the following basic definitions:

**Hazard**: Potential source of harm; deviation from design or operational intent that may constitute or produce a hazard. Hazards are the foci of HAZOP studies. A single hazard may produce multiple forms of harm.

**System**: Subject of a risk assessment, generally a process, product, activity, facility, or logical system.

**Harm**: Physical injury or damage to health or damage to property or the environment. A hazard may impact patient or user safety, employee safety, business, regulatory compliance, the environment, etc.

**Risk**: Combination of probability of occurrence of harm and the severity of the harm, not to be confused with failure mode and critical effect analysis (FMCEA). Risk is not always explicitly identified in HAZOP studies since the core methodology does not require identification (also known as rating) of the probability or severity of harm. Risk assessment teams may choose to rank these factors to further quantify and prioritize risks.

**Node**: A grouping of equipment, lines, vessels, and controls that can be studied by listing all deviations that may occur. The team leader identifies the nodes in advance and confers with the team before they are studied.

**Intent**: How a plant is expected to operate at each node, e.g., supply X amount of Y to pump Z.

**Parameter**: Operational objective, design basis, or physical characteristic of a node such as flow, pressure, temperature, level, vacuum, etc. By using simple guidewords such as NO, MORE, LESS, and by asking questions combining a guideword and parameter such as NO FLOW, a team can determine all types of credible causes in a node.

**Cause**: Event or failure that results in deviation from design intent for a process parameter. "NO FLOW" is a credible cause that describes a blocked pump. After the causes are identified, the consequences are determined.

**Consequence**: Hazard or operability problem resulting from a cause if subsequent events cascade unabated (e.g., trying to start a repaired pump without blocking).

**Safeguard**: Proven safety measure. For example, relief valves must flow enough volume to adequately drop the pressure in all scenarios before it is considered a safeguard.

The above are generally considered default definitions. Readers should understand that different industries and organizations may utilize additional definitions and/or different interpretations. For example, the following definitions may be found in the marine industry:

**Accident:** Event that causes injury, illness, damage, or loss to assets, environment, or third parties.

**Company:** Organization having overall responsibility for a development project and/or operation.

**Contractor:** Third-party organization contracted by a company to perform a specific work scope.

**Guideword**: Term used to facilitate a systematic and structured search for possible deviations from design intent. The list of guidewords is used in conjunction with a list of physical parameters associated with the applicable activity, medium, system conditions, and dynamics.

**Hazard:** Potential source that leads to harm.

**Hazard record:** Record listing potential hazards.

**Incident:** Event or chain of events that may have caused injury, illness, damage, or loss to assets, the environment, or third parties.

**Marine readiness:** Activity normally performed close to the mobilization of an operation.

**Verification:** Process confirming that all adequate preparations and all relevant actions have been completed.

**Risk:** Product of probability of an event and its consequences.

**Risk analysis:** Use of available information to identify hazards and estimate risk.

**Risk assessment:** Overall process of risk analysis and risk evaluation.

**Risk management plan:** Document describing objectives, responsibilities, and activities intended to identify, control, and reduce project risk.

**Risk record:** Record listing potential risks.

**Zero mind-set:** A culture that seeks solutions, designs, and methods in pursuit of a zero accident, incident, or loss philosophy while increasing margins and robustness through application of ALARP.

## Process

### Minimum Requirements

Performance of a HAZOP in any environment requires the following six items at a minimum:

1. HAZARD must be defined. In the most simplistic terms, a hazard is a situation with the potential for an accident that produces undesirable consequences. OSHA's definition in 29 CFR 1910.119 refers to catastrophic hazards involving loss of containment of flammable, combustible, toxic, or highly reactive materials that may affect workers.

2. ACCIDENT or HAZARD SCENARIOS must be developed by an expert team without intimidation. The scenarios should be specific unplanned sequences of events that lead to undesirable consequence.

3. CONSEQUENCES must be discussed and evaluated so that the potential impact of an accident (effects on people, property, or the environment) is evaluated appropriately.

4. OBJECTIVES must be identified clearly so that all parties understand what is at stake. The objectives allow a focus on the identification of all credible accidents (hazard scenarios) that occur frequently and may result in serious injuries to employees or the public, impact the environment, or cause property losses.

5. METHODOLOGY must be defined. Generally, it involves a multidisciplinary team identifying hazards by searching for deviations from design intent via a series of brainstorming meetings. The team should consist of individuals who are familiar with the process to be studied from both technical and operational views. The investigation focuses on deviations from design or operating intent for a process or system, to identify hazard or operability problems by asking questions about one segment of the process at a time. The segments are called nodes.

6. RECOMMENDATIONS for high-priority items must be communicated and acted upon. They must be actionable and completed within a reasonable time.

### Defining Risk

We already discussed the formal definition of risk and its derivatives in Chapter 1. However, when developing a HAZOP, we must keep in mind that risk is the product of probability of occurrence and consequence. Therefore, a HAZOP team must be able to differentiate and define both the consequence and probability categories.

**TABLE 5.1**

A Simple Evaluation Method to Risk

| Personnel Exposure | Delay | Value | Robustness | Suitability |
|---|---|---|---|---|
| Qualification and experience of personnel | Replacement time and cost | Structural strength and robustness | Type of operation Previous experience | Type of operation Previous experience |
| Organization | Repair possibilities | | Installation | Installation |
| Required presence | Number of interfaces and contractors or subcontractors | Operation method | ability Equipment used | ability Equipment used |
| Shift arrangements | Project development period | Novelty and feasibility | Margins, robustness, condition, | Margins, robustness, condition, |
| Deputy and back-up arrangements | Existing field infrastructure | | maintenance | maintenance |
| Overall project particulars | Infrastructure as applicable Handled object | | Previous experience | Previous experience |

**Consequence categories:** consequences should be categorized and each category should have specific criteria in compliance with the health, safety, and environment policies and goals of the organization and governmental regulations if applicable.

**Probability categories:** these should be described qualitatively supplementary guidance from management or organizational policy if needed. Sometimes it is necessary to use surrogate data to develop the required numbers.

**Risk categories:** the consequence and probability categories define risk in a HAZOP. Generally, the risk categories are (1) high, (2) medium, (3) low, and (4) acceptable.

The low-risk category is defined as acceptable subject to application of the principle of ALARP. For the medium and high categories, specified risk identification activities, the principles of ALARP, and risk-reducing activities must be combined to ensure performance of the operation at an acceptable risk level (Det Norske Veritas 2003). The purpose of a risk analysis is to ensure that a hazard has been identified and prioritized. A simplified way of evaluating risk may also be the alternative format shown in Table 5.1. Obviously, the example in the table is generic, but it may help readers understand the definitions and evaluate the ramifications of the results.

## Trigger Events

Typical events that call for HAZOP analysis are:

>Human failure
>
>Equipment, instrument, or component failure
>
>Supply failure

Emergency environment event

Other causes of abnormal operation including instrument disturbance

After an event is triggered, a preliminary or detailed HAZOP analysis may be performed. The minimum requirements are (1) a process flow sheet and (2) process descriptions.

The requirements for a detailed HAZOP are:

Process and instrumentation diagram (P&ID)

Process calculations

Process data sheets

Instrument data sheets

Interlock schedules

Layout requirements

Hazardous area classification

Process description

## Use of Analysis

There are two schools of thought as to when a HAZOP should be performed and how. The first involves a series of activities that include:

1. A high-level review, perhaps HAZID or PHA, very early during the design development.
2. A further HAZID or PHA review after design details are developed. This may also be a what-if analysis or HAZOP.
3. A final design HAZOP should be conducted when the design stage is nearing completion. Note that a detailed P&ID review should be conducted before issuing the drawings as approved for HAZOP.

The second approach is using HAZOP as a design tool early in the design stage but depends on project needs and timeframe. The concept engineering or basic engineering stage may be delayed if this study is included at that stage. In my experience, a HAZOP study is done after all the required documents are available, but it is always better to identify hazards early in the design stage. HAZOP is performed in two stages as shown in Figure 5.1 and Figure 5.2. The first stage takes place after finalization of the process, that is, after the first P&ID is issued. The second stage occurs during detail engineering and should involve all package suppliers.

In the final analysis, a HAZOP is best suited for assessing hazards in facilities, equipment, and processes and is capable of assessing systems from multiple perspectives. The assessment is based on the principle of prevention. For example, in a design situation—an early activity—the focus is on:

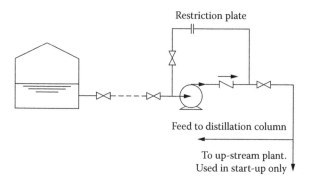

**FIGURE 5.1**
P&ID of feed section of process.

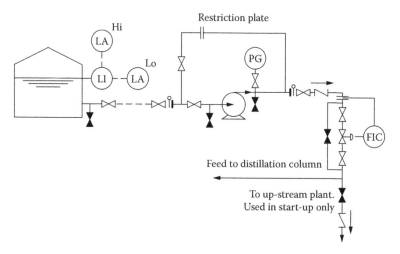

**FIGURE 5.2**
Revised P&ID of feed section of process.

- Assessing system design capability to meet user specifications and safety standards
- Identifying weaknesses in systems, physical, and operational environments
- Assessing environment to ensure the system is appropriately situated, supported, serviced, contained, etc., with operational and procedural controls
- Assessing engineered and automated controls, sequences of operations, procedural controls (human interactions), etc.
- Assessing different operational modes (start-up, stand-by, normal, steady and unsteady states, normal shutdown, emergency shutdown, etc.)

Some issues may arise from the lack of a process to:

- Identify hazards that are difficult to quantify such as:
  - Hazards rooted in human performance and behaviors
  - Hazards that are difficult to detect, analyze, isolate, count, and predict
  - The methodology does not allow the team to explicitly rate or measure deviation, probability of occurrence, severity of impact, or detect hazard
- Identify and evaluate hazards involving interactions of different parts of a system or process
- Identify lack of risk ranking or prioritization capability, in which case the team develops one as required but it may not be consistent with project objectives. Because of the possible inconsistency the team has no way to evaluate effectiveness of proposed or existing safeguards

An assessment may prove worthwhile if based on a free thought-generating process built on the principles of:

Brainstorming methodology

Systematic and comprehensive methodology (logical approach)

A simpler and more intuitive method than other common risk management tools

An interface of HAZOP with other risk management tools such as HACCP

## HAZOP Process

All processes targeted for prevention, including HAZOP analysis, have their own approaches for developing solutions to eliminate or minimize hazards from workplaces. The typical HAZOP is conducted in four steps as summarized in Table 5.2.

### Definition

This phase typically begins with preliminary identification of risk assessment team members. HAZOP is intended to be a cross-functional team effort and relies on subject matter experts (SMEs) from various disciplines with appropriate skills and experience who display intuition and good judgment.

**TABLE 5.2**

HAZOP Steps

| Step | Specific Function |
|---|---|
| Definition | Define scope and objectives |
| | Define responsibilities |
| | Select team |
| Preparation | Plan study |
| | Collect data |
| | Agree on style of recording |
| | Estimate time |
| | Arrange schedule |
| Examination | Divide system into parts |
| | Select part and define design intent |
| | Identify deviations by using guidewords on each element |
| | Identify consequences and causes |
| | Identify whether a significant problem exists |
| | Identify protection, detection, and indicating mechanisms |
| | Identify possible remedial or mitigating measures (optional) |
| | Agree on actions |
| | Repeat for each element and then each part |
| Documentation and follow-up | Record examination |
| | Sign off documentation |
| | Produce report of study |
| | Follow up to ensure actions are implemented |
| | Restudy any parts of system if necessary |
| | Produce final output report |

SMEs should be carefully chosen to cover all relevant experiences and functions. All meetings should be conducted in a positive atmosphere, and all members should be able to contribute their ideas without fear of retaliation. During this phase, the risk assessment team must carefully identify and agree on the scope in order to focus their efforts. They must define the project boundaries, key interfaces, and important assumptions that determine the direction of the assessment.

## Preparation

This phase typically includes:

1. Identifying and locating supporting data and information
2. Identifying the audience and users of the study outputs
3. Preparing appropriate and applicable project management issues such as: scheduling meetings and transcribing proceedings

4. Reaching consensus on template format for recording study outputs
5. Reaching consensus on HAZOP guidewords for the project HAZOP guidewords are key supporting elements in the execution of a HAZOP analysis

According to IEC Standard 61882, the identification of deviations from design intent is achieved by a questioning process using predetermined guidewords. The role of a guideword is to stimulate imaginative thinking, focus a study, and elicit ideas and discussion. In essence the HAZOP guidewords can be used to stimulate brainstorming of potential risks and utilize additional risk assessment tools.

Risk assessment teams are responsible for identifying the guidewords that best suit the scope and problem statement for their analysis. Some common HAZOP guidewords are listed below. HAZOP guidewords work by providing a systematic and consistent means of brainstorming potential deviations from standard operation. After the HAZOP guidewords are selected, the examination phase may begin. The following list shows how guidewords may be used to brainstorm deviations related to detergent control for a cleaning operation. Others can be crafted as needed.

No or not—no detergent added

Part of—critical detergent component (e.g., surfactant) omitted

More—too much detergent added (difficult to rinse); detergent solution concentration too high

Reverse—detergent contaminated with harmful hazard

Other than—wrong detergent used

Less—too little detergent added (soil not effectively removed); detergent solution concentration too low

Early—detergent added too early (e.g., during pre-rinse)

Late—detergent added too late in cleaning cycle

As well as: satisfactory or good way as before detergent was added

Before: condition before the detergent was added

After: Condition after the detergent was added

Reverse (of intent): results were opposite of what was expected

## Examination

This phase begins with identification of all elements, parts, or steps of the system or process to be examined. A process flow diagram is a good tool for this purpose. It allows physical systems to be broken down into smaller parts as necessary. Processes may be broken down into discrete steps or phases. Similar parts or steps may be grouped to facilitate assessment.

The HAZOP guidewords are then applied to each element to achieve a systematic and thorough search for deviations. It must be noted that not all combinations of guidewords and elements will yield credible deviation possibilities. As a general rule, all reasonable use and misuse conditions expected by the user should be identified and subsequently challenged to determine whether they are credible and should be assessed further. There is no need to explicitly document combinations of elements and guidewords that fail to yield credible deviations.

## Documentation and Follow-Up

The documentation of HAZOP analyses is often facilitated by utilizing a template recording form as detailed in IEC Standard 61882. A typical form is shown in Table 5.3. Risk assessment teams may modify the template as necessary based on factors such as:

Regulatory requirements

Need for more explicit risk rating or prioritization (e.g., rating deviation probabilities, severities, and/or detection)

Company documentation policies

Needs for traceability or audit readiness

Other factors

Finally, before the detailed HAZOP takes place, the process (or pipeline is relevant) and instrumentation diagram (P&ID) must be developed during the last stage of process design. A P&ID is a schematic of functional relationships of piping, instrumentation, and equipment components, without which a process cannot be designed adequately. A P&ID:

Represents the last step in process design.

Shows all piping, including the sequences of branches, reducers, valves, equipment, instrumentation, and control interlocks.

Is normally developed from a process flow diagram (PFD).

Is used to operate the process system.

**TABLE 5.3**

HAZOP Recording Form

| Process | Deviation | Cause | Consequence | Safeguards | Action |
|---------|-----------|-------|-------------|------------|--------|
|         |           |       |             |            |        |
|         |           |       |             |            |        |
|         |           |       |             |            |        |
|         |           |       |             |            |        |

The development of a P&ID will follow a normal standard procedure and include:

Basic process control system functioning as a closed loop for maintaining processes within a defined operating region

Alarm system to bring unusual situations to the attention of the person monitoring the process

Safety interlock system to stop the operation or part of it during emergencies

Relief system to divert material safely during emergencies

A P&ID should cover every mechanical aspect of a plant:

Instrumentation and designations

Mechanical equipment with names and numbers

All valves and their identifications

Process piping, sizes, and identification data

Miscellaneous vents, drains, special fittings, sampling lines, reducers, increasers, and swagers

Permanent start-up and flush lines

Flow directions

Interconnections references

Control inputs, outputs, and interlocks

Interfaces for class changes

Seismic category

Quality level

Annunciation inputs

Computer control system input

Vendor and contractor interfaces

Identification of components and subsystems delivered by others

Intended physical sequence of equipment

On the other hand, a P&ID should *not* include:

Instrument root valves

Control relays

Manual switches

Equipment rating or capacity

Primary instrument tubing and valves

Pressure temperature and flow data

Elbows, tees, and similar standard fittings

Extensive explanatory notes

## Detailed Analysis

The last section provided a general overview of HAZOP. This section will develop the process with more detail. A HAZOP examination systematically questions every part of a process or operation to discover qualitatively how deviations from normal operation can occur and whether further protective measures, altered operating procedures, or design changes are required. Figure 5.3 shows the HAZOP procedure in a flowchart characterization.

The examination procedure uses a full description of the process and will almost invariably include a P&ID or equivalent. It systematically questions

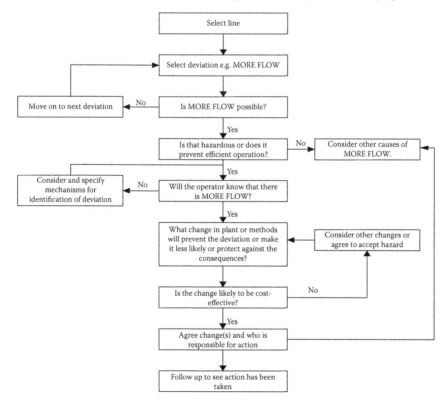

**FIGURE 5.3**
HAZOP procedure flow. (*Source:* HIPAP 8. 2011. With permission)

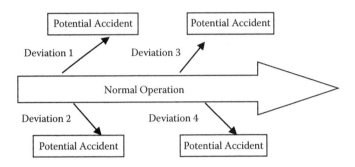

**FIGURE 5.4**
Operation with deviations.

every part of the process to determine how deviations from the intent of the design can occur and determine whether they will give rise to hazards.

The questioning is sequentially focused around guidewords are derived from a team discussion using Q&A or other investigative techniques. The guidewords ensure that the questions posed to test the integrity of each part of the design will explore every conceivable way in which operation could deviate from the design intent.

Some of the causes may be so unlikely that the derived consequences will be rejected as not meaningful. Some of the consequences may be trivial and need to be considered no further. However, some deviations have conceivable causes and potentially serious consequences (see Figure 5.4). The potential problems are then noted for remedial action. The immediate solution to a problem may not be obvious and may need further consideration by a team member or perhaps a specialist. All decisions must be recorded.

Traditionally, recorders were designated to take notes. However, in the age of technology, software may be used to assist in recording HAZOP proceedings but should not be considered as a replacement for an experienced chairperson and secretary. The main advantage of the software approach is its systematic thoroughness in failure identification. The method may be used at the design stage when plant alterations or extensions are planned or may be applied to an existing facility.

## Sequence of Examination

The logical sequence in conducting a HAZOP is determining:

| | |
|---|---|
| Intent | Consequences (hazards and operating difficulties) |
| Deviation | Safeguards |
| Causes | Corrective actions |

Typically, a member of the team outlines the purpose of a chosen line of the process and how it is expected to operate. The various guidewords such as MORE are selected in turn. The team then considers what issues could cause the deviation.

The next step is considering the results of a deviation, for example, the creation of a hazardous situation or operational difficulties. When the considered events are credible and the effects significant, existing safeguards should be evaluated and a decision made as to what additional measures could be required to eliminate the identified cause. A more detailed reliability analysis such as risk or consequence quantification may be required to determine whether the frequency or outcome of an event is serious enough to justify major design changes.

## Deviations from Design Intent

The design intent is the reason a plant, equipment, or process exists. Intent is fundamental to a HAZOP analysis because any deviation based on intent must be considered and handled. The primary intent is critical to understanding the overall main design intent. In many cases, an important secondary intent is ensuring operation in the safest and most efficient manner (Lihou n.d.).

Both the primary and secondary intents of the design are important factors and should be included in the HAZOP to achieve the desired goals. Each item of equipment—each pump and length of pipework—must consistently function in a specific manner. This consistent function may be classified as the design intent for that item. To demonstrate this concept, we adopt a simple illustration of a cooling water facility (Lihou n.d.). See Figure 5.5. The design would certainly require cooling water circuit pipework and a pump.

A simple statement of the design intent of this small section of the plant would be "continuously circulate cooling water at an initial temperature of x°C and at a rate of x liters per hour." It is usually at this low level of design

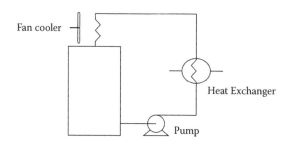

**FIGURE 5.5**
Cooling water facility. (*Source:* Lihou, M. http://www.lihoutech.com/hzp1frm.htm)

intent that a HAZOP study is directed. The *deviation* term now becomes easier to understand. A deviation or departure from the design intent in the case of the cooling facility would be a cessation of circulation or the water at a too-high initial temperature. Note the difference between a *deviation* and its *cause*. In this case, failure of the pump would be a cause, not a deviation.

## Details of Study Procedure

The study of each section of a plant generally follows the following sequence:

1. The process designer briefly outlines the broad purpose of the section of the design under study and displays the P&ID (or equivalent) where it can be readily seen by all team members.

2. General questions about the scope and intent of the design are discussed.

3. The relevant part for study is selected, usually one in which a major material flow enters that section of the plant. The part or item is highlighted on the P&ID with dotted lines using a transparent pale colored felt pen (see Figure I.3 in the Introduction to this book).

4. The process designer explains in detail the purpose, design features, operating conditions, fittings, instrumentation, protective systems, and details of the processes immediately upstream and downstream if relevant.

5. Any general questions about the part or item are discussed.

6. The detailed line-by-line study commences. The discussion leader reviews the guidewords chosen as relevant. Each guideword such as HIGH FLOW identifies a deviation from normal operating conditions to prompt discussion of the possible causes and effects of flow at an undesirably high rate. If, in the opinion of the study team, the combination of the consequences and the likelihood of occurrence are sufficient to warrant action, the combination is regarded as a problem and recorded as such. If the existing safeguards are deemed sufficient, no further action is required. For major risk areas, the need for action may be assessed quantitatively using techniques such as hazard analysis (HAZAN) or reliability analysis. For less critical risks, assessment is usually based on experience and judgment. The person responsible for defining the corrective action is also nominated.

7. The main aim of the meeting is to find problems needing solutions rather than finding solutions. The group should not be tied down by trying to resolve problems. It is better to proceed with the study, deferring consideration of the unsolved problems to a later date.

8. When a guideword requires no more consideration, the chairperson refers the team to the next guideword.

9. Discussion of each guideword is confined to the section, part, or item marked, the processes at each end and any related equipment such as pumps or heat exchangers. Any changes agreed at the meeting are recorded and if appropriate marked on the P&ID or layout with red pen.

10. When all guidewords have been covered, the line is fully highlighted to show that it has been completed, and the next line is chosen.

11. When all the lines in a plant subsection have been reviewed, additional guidewords are used for an overview of the P&ID.

## Effectiveness Factors

The effectiveness of a HAZOP will depend on several things, including:

1. The accuracy of information (including P&IDs) available to the team; information should be complete and up to date.

2. The skills and insights of team members.

3. The ability of the team to use a systematic method to identify deviations.

4. Maintenance of a sense of proportion in assessing the seriousness of a hazard and the expenditure of resources to reduce its likelihood.

5. The competence of the chairperson in ensuring the team rigorously follows sound procedures.

The key elements of a HAZOP are (1) the team; (2) a full description of the process to be examined; (3) relevant guidewords; (4) conditions conducive to brainstorming; (5) recording of meetings; and (6) a follow-up plan.

## Team

The HAZOP team will typically consist of five to nine people. A team should have an odd number of members to eliminate the possible ties. Team members should have a range of relevant skills to ensure all aspects of the plant and its operations are covered. Engineering disciplines, management, and plant operating staff should be represented. This will prevent possible events from being overlooked through lack of expertise and awareness. It is essential that the chairperson is experienced in HAZOP techniques to ensure that the team follows the procedure without diverging or taking shortcuts.

Where a HAZOP is required as a condition of development consent, the name of the chairperson is typically required to be submitted to the appropriate regulators for approval before commencement of the HAZOP. It is also important that the study team be very familiar with the information in the plant P&ID or other description of the process considered. For an existing plant, the group should include experienced operational and maintenance staff. In defining a team, important selection criteria should be considered.

### Team Leader (Chairperson)

The HAZOP team leader works with the project coordinator in defining the scope of the analysis and selection of team members. He or she directs the team members in gathering of process safety information before the start of the study and also plans the study with the project coordinator and schedules team meetings. The leader directs the team in the analysis of the selected process, keeping members focused on discovering hazards associated with the process and directs the team scribe in recording the findings. He or she ensures that the analysis thoroughly covers the process defined at the start of the exercise and that it is completed in the time allotted during the planning stage. After the analysis is complete, he or she writes a report detailing the findings and recommendations of the team and submits it to management. The last task is fielding inquiries about the project recommendations.

### Engineers

The engineering experts assigned to the HAZOP may include any combination of project engineers, machinery engineers, instrument engineers, electrical engineers, mechanical engineers, safety engineers, quality assurance engineers, maintenance engineers or technicians, and corrosion and materials engineers. These individuals should provide expertise in their respective disciplines as it applies to the process under study. They are also responsible for attending the initial hazard analysis meeting. They must be available to the team as required with the understanding that the team leader will give adequate advance notice to the experts when possible. The experts must provide documentation of all existing safeguards and procedures. A HAZOP team assigned to consider a new chemical plant could include:

- Chairperson: an independent person who has sound knowledge and experience of HAZOP techniques. Some understanding of the proposed plant design would also be beneficial.
- Design engineer: the project design engineer, usually mechanical, who has been involved in the design and is concerned with project cost.
- Process engineer: usually the chemical engineer responsible for the process flow diagram and development of the P&IDs.

- Electrical engineer: usually the engineer responsible for design of the electrical systems in the plant.
- Instrument engineer: the instrument engineer who designed and selected the plant control systems.
- Operations manager: preferably the person who will be in charge of the plant when it moves to the commissioning and operating stages.

A team with a narrower range of skills is unlikely to be able to satisfactorily conduct a HAZOP of this nature. Of course, other skills may be needed. For example, if a plant uses a new chemical process, a research chemist may be required. Including an experienced supervisor or operator on the team is also often appropriate, especially one from a similar plant already in operation. It is imperative that at least one member of the team has sufficient authority to make decisions affecting design or operation of the facility, including decisions involving substantial additional costs.

## Description of Process

A full description of the process is needed to guide the HAZOP team. This presupposes that a good understanding of the process exists. It also presupposes that the appropriate and applicable individuals form the team. In the case of conventional chemical plants, detailed P&IDs should be available. At least one member of the HAZOP team should be familiar with the applicable diagrams and all instrumentation they represent. If a plant is very complex or large, it may be split into a smaller number of units (nodes) to be analyzed at separate HAZOP meetings.

In addition to P&IDs, physical or computer-generated models of the plant or photographs of similar existing plants may also be utilized. The additional documents greatly help the team visualize potential incidents, especially those caused by human error. An inspection by the HAZOP team of a similar plant already in operation before commencement of the HAZOP would be highly beneficial. If a similar plant is in operation, the team should review past incidents. Key information that may be required during the HAZOP should also be readily available and should include:

- Layout drawings
- Hazardous area drawings
- Material safety data sheets
- Relevant codes or standards
- Operating manual for existing plant
- Outline operating procedures for new plant

When carrying out a HAZOP on a facility for which traditional P&IDs are not appropriate, it may be more suitable to use alternative visualization and diagrammatic techniques such as plan and section drawings, layout drawings, or photographs. A decision as to the medium to be used should be made well before the HAZOP commences.

In batch processes, additional complexities are introduced into the HAZOP because of the time-dependent nature of the system components. It is strongly recommended that the chairperson be knowledgeable about the process at hand and also experienced in batch process HAZOPs.

### Relevant Guidewords

A set of guidewords relevant to the operation should be chosen, studied, and systematically applied to all parts of the operation. This may entail application of the guidewords to each process line within a P&ID or by following each stage of an operation from start to finish. Table 5.4 shows examples of guidewords and variations. The choice of suitable guidewords will strongly impact the success of a HAZOP in detecting design faults and operability problems.

In addition to reviewing and analyzing normal operations, a HAZOP should also consider conditions during plant start-up, shut-down, and all applicable modifications. Human response time and the possibility of inappropriate action by an operator or supervisor should also be considered. If such errors are possible, it is suggested that a mistake-proofing methodology be designed and implemented to avoid human errors.

### Point of Reference Concept

The concept of point of reference (POR) is useful when defining nodes, evaluating deviations and performing a HAZOP on an individual node. As an example, let us look at a flash drum—a vessel into which flows a mixture of liquid and vapor. The goal is to separate the vapor and liquid. For design calculations, it is normally assumed that the vapor and liquid are in equilibrium and that the vessel is adiabatic (no heat lost or gained). The goal is to simultaneously satisfy a material balance, heat balance, and equilibrium. This can be done by the use of a separator or a tank (McGraw-Hill 2003). In this case, the node consists of the flash drum and liquid product piping up to the flange on a product storage tank. If a no-flow deviation is proposed, a dilemma becomes apparent in discussion of the causes of no flow.

If a no-flow situation arises because of a pipe rupture at the flange connection on the flash drum, the rupture is the cause. The no-flow term is ambiguous because there is flow out of the flash drum but not through the piping to the storage tank. Therefore, a POR should be clearly established when the node is defined. It is recommended to always establish a POR at the downstream end of a node.

**TABLE 5.4**

Typical Guidewords

| Guideword | Meaning | Comments |
|---|---|---|
| NO | Complete negation | For example, of intention |
| NO | Forward flow | When there should be |
| MORE | Quantitative increase | More of a relevant physical property than there should be (e.g., high flow, temperature, pressure, viscosity; actions (heat and chemical reaction) |
| LESS | Quantitative decrease | Less of a relevant physical property, etc. |
| AS WELL AS | Quantitative increase | All design and operating intentions achieved together with some addition (e.g., impurities, extra phase) |
| PART OF | Quantitative decrease | Only some intentions achieved |
| REVERSE | Opposite of intention | Reverse flow or chemical reaction (e,g., inject acid instead of alkali in pH control) |
| OTHER THAN | Complete substitution or miscellaneous | No part of original intention; different result achieved; start-up, shutdown, alternative mode of operation, catalyst change, corrosion, etc. |

*Guidewords for Line-by-Line Analysis*

| | |
|---|---|
| FLOW | HIGH, LOW, ZERO, REVERSE |
| LEVEL | HIGH, LOW, EMPTY |
| PRESSURE | HIGH, LOW |
| TEMPERATURE | HIGH, LOW |
| IMPURITIES | GASEOUS, LIQUID, SOLID |
| CHANGE IN COMPONENTS | |
| CHANGE IN COMPOSITION | |
| CHANGES IN CONCENTRATION | |

*Two-Phase Flow Reactions*

| | |
|---|---|
| TESTING | Equipment or product |
| PLANT EQUIPMENT | Operable or maintainable |
| INSTRUMENTS | Sufficient, excess, location |
| ELECTRICAL EQUIPMENT | Area isolation, grounding |

*Continued*

**TABLE 5.4 (Continued)**

Typical Guidewords

| Guideword | Meaning | Comments |
|---|---|---|
| *Overview (after Line-by-Line Analysis)* | | |
| TOXICITY | | |
| COMMISSIONING | | |
| START UP | | |
| INTERLOCKS | | |
| BREAKDOWN | | Including services and computer failure |
| SHUTDOWN | | Purging, isolation |
| EFFLUENT | | Gaseous, liquid, solid |
| NOISE | | |
| TESTING | | Product or equipment |
| FIRE AND EXPLOSION | | |
| QUALITY AND CONSISTENCY | | |
| OUTPUT | | Reliability, bottlenecks |
| EFFICIENCY, LOSSES | | |
| SIMPLICITY | | |
| SERVICES REQUIRED | | |
| MATERIALS OF CONSTRUCTION | | Vessels, pipelines, pumps, etc. |
| SAFETY EQUIPMENT | | Personal, fire detection and fighting, means of escape |

## Screening for Causes of Deviations

It is necessary to list causes of deviations thoroughly. A deviation is considered realistic if it presents sufficient causes to be likely. Only credible causes should be listed. Team judgment is used to decide whether to include events with very low probabilities. However, good judgment must be used in determining what events have low probabilities of occurring so that credible causes are not overlooked. The three basic types of causes and examples are:

1. Human errors: acts of omission or commission by an operator, designer, contractor, or other person who creates a hazard that could possibly result in a release of hazardous or flammable material.

2. Equipment failure: a mechanical, structural, or operating failure results in the release of a hazardous or flammable material.

3. External event: an item outside the relevant unit affects the operation of the unit to the extent that the release of hazardous or flammable materials is possible.

External events include malfunctions of adjacent units affecting the safe operation of the unit or node under study, loss of utilities, and impacts of weather and seismic activity.

The level of detail required in describing causes of a deviation depends on whether the cause of the upset occurs in inside or outside the node. For example, suppose that a drum includes a level controller as part of the node and the control valve closes, creating a high-level condition. Since the valve and controller are parts of the node, the causes should be stated in more detail.

The valve may close because the wrong set point was input by an operator (human error); the valve may remain closed due to mechanical failure; or the valve may close due to loss of instrument air to the unit (external event). If the level controller is located outside the node under study, a sufficient statement is "Level control valve LV-XXXX closes."

When the team reaches the node in which the level controller is located, more detail about causes can be listed. A good technique is to *screen for causes of deviations based on human error, equipment failure, and applicable external events.*

### Consequences and Safeguards

One purpose of a HAZOP is identification of scenarios that could lead to the release of hazardous or flammable material into the atmosphere, exposing workers to injury. In order to make this determination, it is always necessary to examine as exactly as possible all consequences of any credible causes of a release identified by the team. This will serve a twofold purpose. First, it will help the team devise a risk ranking in a HAZOP that reveals multiple hazards. This will allow the team to set priorities for addressing the hazards. Second, it will help the team determine whether a specific deviation results in an operability problem or hazard.

If the team concludes from the consequences that a particular deviation results in an operability problem only, the discussion should end and the team should move on to the next cause, deviation, or node. If the team determines that the cause may lead to the release of hazardous or flammable material, safeguards should be identified. Safeguards should be included whenever a team determines that a combination of cause and consequence presents a credible process hazard. What constitutes a safeguard can be summarized by utilizing:

1. Systems, engineered designs, and written procedures designed to prevent a catastrophic release of a hazardous or flammable material.
2. Systems designed to detect and give early warning following the initiating cause of a release of a hazardous or flammable material.
3. Systems or written procedures that mitigate the consequences of a release of a hazardous or flammable material.

The team should use care when listing safeguards. Hazards analysis requires evaluation of the consequences of failures of engineering and administrative controls and a careful determination of whether items are genuine safeguards must be made. In addition, the team should consider realistic multiple failures and simultaneous events when determining whether a safeguard will actually function as planned in the event of an occurrence.

### Deriving Recommendations (Closure)

Recommendations are made when the safeguards for a given hazard scenario, as judged by an assessment of the risk of the scenario, are inadequate to protect against a hazard. Action items require recommendations from the relevant individual or department. If software is an action item, information needs are identified as recommendations for follow-up by one of the team members. The following guidelines are suggested for the implementation of recommendations. For obvious reasons, the time limits cited may be adjusted to reflect the organization's needs as well as the urgency and complexity of the recommended solution:

> High priority action items should be resolved within 4 months.
>
> Medium priority action items should be resolved within 4 to 6 months.
>
> Lower priority action items should be resolved after medium priority items are handled.

The facility's safety coordinator should review all recommendations of a HAZOP to determine relative priorities and a schedule of implementation. After each recommendation has been reviewed, the resolution should be recorded in a tracking document such as a spreadsheet and kept on file. Recommendations include design, operating, or maintenance changes that reduce or eliminate deviations, causes, and/or consequences. Recommendations identified in a HAZOP are considered preliminary in nature. Additional information or study can also be recommended.

### Conditions Conducive to Brainstorming

The HAZOP should be performed under conditions conducive to brainstorming. The team should meet in an area that is free from interruptions and includes facilities for displaying diagrams. Whiteboards or other recording media should be available. The minutes should be recorded completely and clearly, preferably not by the chairperson.

### Meeting Records

One approach to record keeping is to record only key findings (reporting by exception). A second technique is recording all issues. Experience has shown

**TABLE 5.5**

HAZOP Meeting Record

| Project: | | | | Node: | | Page: | |
|---|---|---|---|---|---|---|---|
| Description: | | | | | | Date: | |
| | | | | | | Drawing/ Revision No.: | |
| Guideword | Cause | Consequence | Safeguard | Rec # | Recommendation | Indiv | Action |
| | | | | | | | |
| | | | | | | | |
| | | | | | | | |

that reporting by exception can be adopted in most cases because it minimizes the clerical work and focuses on issues that need attention. It is important, however, that the recording of safeguard information is retained, even when no further action is required. Such records ensure that safeguards will not be removed through ignorance after the HAZOP is complete.

Table 5.5 is a HAZOP meeting record. It is not intended to be definitive; it depicts one suitable way of recording results. It is generally acknowledged that a HAZOP process becomes tedious over an extended period. Sessions should he kept to half a day if possible if the exercise is likely to extend over several clays. It is also important to ensure maximum participation in the study by each team member. Attendance at all sessions should be mandatory. Care should be exercised to provide physical surroundings conducive to participation.

Huge numbers of records may be generated by a HAZOP. If this is the case, only records for which incidents may occur or where it is *not* obvious that such incidents cannot occur need be included with the report. A comprehensive set of all records generated by the HAZOP should be kept for use by the company and the department where the HAZOP took place.

The success of the HAZOP methodology when applied to continuous and to batch processing operations is well proven. The technique with modifications can also be applied to automotive production, finance, healthcare, and many more industries.

### Meeting Questions

In any meeting dealing with risk and or hazards, many questions arise. The following questions must be answered to ensure a complete discussion and substantial corrective action for the risks and/or hazards identified:

- Is there management support?
- Are supervisors and employees trained on using HAZOP?

- Is there a written program?
- Is there a program director?
- Who has access to the completed HAZOP?
- How is the completed HAZOP used?
- What techniques are used to develop:
  - One-on-one observations?
  - Group participation?
  - Recall steps?
  - Absentee issues?

**Follow-Up**

The fact that a HAZOP analysis is conducted to eliminate or minimize hazards cannot be overstated. Arriving at a solution is only a partial answer to a concern. The other part is to make sure the solution is workable within the time constraints and ensure proper follow-up to demonstrate that the integrity of the solution is what it should be over the short and the long terms defined by the team.

**Computer HAZOP (CHAZOP)**

The use of electrical, electronic, or programmable electronic (E/E/PE) systems in safety-related applications is steadily growing. The method is applicable to computer-based instrumentation and control and safety-related functional applications in modern chemical plants and related industrial situations. Difficulties arising from malfunctions of such systems are also increasing, particularly as experience with such systems flags new problems that were not encountered in older plant designs. The interface with modern electronic controls and protective systems remains a potential weakness in the overall reliability of these systems.

The E/E/PE systems relating to the operations function of the plant may be tested regularly "on the run." However, this may not be true for safety-related systems intended to perform infrequently—in the event of a failure or dangerous situation. Dangerous situations can arise from:

- Inadequate specification of the hardware and software requirements of a functional safety system at the design stage
- Inadequate consideration of modifications to software and hardware
- Common cause (system) failures
- Human errors
- Random hardware faults

- Extreme variations in surrounding conditions (electromagnetic emissions, temperature, vibration)
- Extreme variations in supply systems (low or high supply voltage, loss of air pressure for emergency shutdown, voltage spikes on resumption after power outage)

A hazard analysis determines whether functional safety is necessary to ensure adequate protection. Functional safety is part of an overall safety plan that depends on correct operation of a system or device in response to its inputs. An overpressure protection system using a pressure sensor to initiate the opening of a relief device before dangerous high pressures are reached is an example of functional safety. Two types of requirements are necessary to achieve functional safety: (1) safety function requirements (what the function does in relation to design intent); and (2) safety integrity requirements (likelihood that a safety function will perform satisfactorily).

The safety function requirements are derived from the hazard analysis. The safety integrity requirements result from the risk assessment. The HAZOP or CHAZOP should review the safety-related systems that must operate satisfactorily to achieve a safe outcome in the event of an incident with potential to produce a dangerous failure. The aim should be to ensure that the safety integrity of the safety function is sufficient to prevent exposure to an unacceptable risk associated with a hazardous event.

The importance of E/E/PE systems has increased in recent years, particularly for computer control and software logic interlocks. If a computer and instrumentation system are sufficiently complex for the facility, it may be useful to consider them in a separate HAZOP (computer-based HAZOP or CHAZOP)—both control and protective—or as discrete components of a more general HAZOP.

Modern plants almost invariably include E/E/PE systems that present different spectra of failure modes from those encountered in conventional HAZOPs. The flexibility of E/E/PE systems that offers the capability to control several complex operations can also provide more possibilities for errors than conventional control systems. The likelihood of common mode failures increases with such systems; for example, the failure of a single input/output (I/O) card may cause the losses of several control and information channels. A CHAZOP will highlight such issues and lead to corrective solutions such as employing two independent systems or hardwiring key control circuits.

A discrete study of control systems and safety-related components can be particularly valuable where instrumentation has been designed and installed as a package unit by a contractor. It allows a HAZOP team to gain an understanding of the system. Treating these items as discrete components of a HAZOP allows the operator–computer interaction to be examined. However, plant management should not forget that the overall plant HAZOP will not be complete until the E/E/PE systems have been reviewed by CHAZOP or equivalent technique.

These issues can be reviewed by other disciplined techniques along the lines of HAZOP. Clearly, such techniques must be adapted and refined appropriately to be suitable for a particular system. As with a HAZOP, a CHAZOP assessment team should record all the hazards identified by the computer technique and provide recommendation for possible improvement.

### Advantages and Disadvantages

- Advantage; methodical study of errors within the software by systematically applying a set of guidewords to the software and process control systems.
- Disadvantages:
  - Can be expensive and time consuming. A complex process will require evaluation of a large number of computer systems and components.
  - Can only be applied to software and control systems. The technique is specifically designed to identify hazards in computer systems and cannot be easily applied beyond that area.

### Human Factors HAZOP

An expanded approach devised by Pitblado et al. (1989) is to conduct a multi-tiered HAZOP study in which a conventional HAZOP forms only the first tier. A computer system HAZOP (CHAZOP) becomes the next stage; A human factors HAZOP constitutes the third and final stage. Different guidewords are used at each tier. We believe that with appropriately modified guidewords, the HAZOP technique can be applied to situations that are not strictly process oriented. Even if no strictly disciplined technique is employed, a searching study of materials handling and warehousing and other operations would benefit from the team study approach of HAZOP.

---

## Report

At the completion of a HAZOP analysis, a full report must be issued. The minimum requirements for a report are explained below.

### Study Title Page

The study title must be shown on the cover and on a separate title sheet. The title should clearly and unambiguously identify the facility covered by the study. The title page should also show the type of operation (proposed or existing) and its location. The title sheet should specify who authorized the report and the date it was authorized. The chairperson and organization she or he represents should also be noted.

## Table of Contents

A table of contents must be included at the beginning of the report. It should list report sections or contents and also list figures, tables, and appendices.

## Glossary and Abbreviations

A glossary of special terms or titles and a list of abbreviations must be included to ensure that the report can be easily understood.

## Aim

The report must provide sufficient information about each element and adequate cross references. Any section read alone or in sequence with other sections should allow and assessment of the adequacy of the HAZOP study.

## Guidewords

The guidewords used to identify possible deviations of operations must be listed. Furthermore, explanations of all specialized words that apply to the facility should also be given. See Table 5.6.

## Summary of Main Findings and Recommendations

A summary must briefly outline the nature of the proposal or existing facility and the scope of the report. A list of the main conclusions and recommendations arising from the HAZOP must be presented. It is also useful to include the implementation timetable.

## Scope of Report

This section should briefly describe the aims and purpose of the study. For example, is the study intended to satisfy conditions of development consent or as a company initiative as a component of a safety upgrade? Does the study cover a new development or a modification or extension of an existing facility. Reference must be made to other relevant safety related studies completed or under preparation.

## Description of Facility

This section should provide an overview of the site, plant, and materials stored and used there. If such information is already available in an

**TABLE 5.6**

Guidewords and Parameters

| Guideword | Parameter |
|-----------|-----------|
| No | Flow |
| Reverse | |
| More | |
| Less | |
| More | Pressure |
| Less | |
| More | Temperature |
| Less | |
| More | Level |
| Less | |
| No | |
| More | Phase |
| Less | |
| More | Composition |
| Less | |
| Other | Start-up |
| | Shutdown |
| | Commissioning |
| | Relief and blow-down |
| | Draining |
| | Venting |
| | Isolation |
| | Purging |
| | Sample points |
| | Instruments |
| | Maintenance access |
| | Construction materials |
| | Static electricity |

*Note:* Table 5.4 explains these guide-words and parameters.

environmental impact statement (EIS), hazard analysis, or other document, clear cross references to such documents or their inclusion as appendices will suffice. The description should include:

1. Sketch of the site location with identification of adjacent land uses.
2. Schematic diagram of the plant under study along with brief descriptions of all process steps. The locations and natures of raw materials, product storage, and loading and unloading facilities must also be

shown. The plant does not have to be described in detail, although some process conditions such as pressure in pressurized vessels may be necessary to explain some operations.

3. Clearly identified P&IDs with plant and line numbers as used in the HAZOP. Instrumentation and equipment symbols should be explained. Alternatives media (photographs, plans, etc.) must carry appropriate identification. If a large number of P&IDs are involved in the study, only those relevant to the recommendations should be appended to the report.

## Team Members

All HAZOP participants and their affiliations and positions should be noted. Their responsibilities, qualifications, and relevant experience should also be shown. The chairperson and the secretary of the group should be identified. The dates of the meetings and their duration should be listed. If some members did not attend all meetings, the extent of their participation should be indicated. Special visitors and occasional members must be listed in a similar manner and the reasons for their attendance detailed. For example, specialist instrumentation engineers and consultants may be required to attend certain sessions to overcome specific design problems.

## Methodology

The general approach must he briefly outlined. Any changes to the accepted standard method for a HAZOP must be detailed and explained.

## Overview

This section must outline the conditions and situations likely to result in a potentially hazardous outcome considered in the HAZOP (following line-by-line analysis) for each P&ID or section, including overview issues, such as:

- Start-up procedures
- Emergency shutdown procedures
- Alarm and instrumentation trip testing
- Pre-commissioning operator training
- Plant protection systems
- Service failure
- Breakdowns
- Effluents (gas, liquid, solid)
- Noise

Any issues raised and considered necessary for review outside the HAZOP must be detailed. See Table 5.3 for some typical guidewords.

### Analysis of Main Findings

The criteria used to determine whether action was required should be noted in this section. Also, the results showing deviations, consequences, and required actions required, must be explained. Events that required no actions should also be listed, as should events with minor consequences or conditions not requiring risk analysis. The decisions made after such further analyses must also be noted along with alternative actions generated and considered.

### Findings

This section highlights items that are potentially hazardous to plant personnel, the public, or the environment or have the potential to jeopardize plant operability. The section should include a clear statement of commitment to modify the design or operational procedures in accordance with the required actions and a timetable for implementation. Justifications for not taking certain actions should also be explained. The current status of the recommended actions at the time of the report should also be noted along with the names and titles of persons responsible for their implementation.

## Review

In a very broad sense, a HAZOP review is meant to identify actions required to alleviate or remove potential hazards or operability problems revealed by the study. Proper recording and reporting of the HAZOP review discussion is an integral part of a HAZOP review. The scope of the HAZOP review must be clearly stated in the info pack document (see below). As a guideline, items such as positioning of safety showers, valve accessibility, handling of process and waste chemicals are not included in a HAZOP review. The intent is to establish guidelines for a HAZOP review.

### Input Documents

The following input documents are required for a HAZOP review:

- HAZOP review information package (info pack)
- Appropriately sized P&IDs "approved for design"

- Appropriately sized plot plans "approved for design"
- Appropriately sized process safety flow schematics if available at the time of HAZOP
- Soft (preferably .pdf) files of the review drawings and documents for display on a projection screen at the HAZOP review venue

### Review Information Pack

The HAZOP review info pack is basically a compilation of all the information needed for a HAZOP review and provided to each participant a few days or more before the review. The following should be included in an info pack:

- A Word document containing the following information in sequence:
  - Summary: brief overview of the project to be studied.
  - List of abbreviations and definitions used in the Word document.
  - Introduction noting the purpose of the project and the background or a brief overview of the project.
  - Process description or narrative of the process, including the individual unit operations and their sequences.
  - Scope information, including a list of drawings and documents to be referenced in the review.
  - Node details: a narrative of the P&ID nodes to be reviewed. The nodes should be highlighted in the P&IDs to be used for the review and included in the info pack.
  - Appendices may used to list additional documents and drawings for the review. Appendices are optional and may be used if applicable in each individual case.
- For ease of distribution of the info pack, appropriate size drawings of P&IDs, plot plans, and other drawings listed in the scope section must be attached to the main Word document. An appropriate number of copies of the compiled info pack, including the Word document and drawing attachments must be made to distribute to all participants in the HAZOP review and then distributed to the individual study participants. The preparation and distribution of the info packs are the responsibilities of the project engineer for the system or facility to be subjected to HAZOP analysis.

### Review Team Composition

To complete an effective HAZOP, a balanced multi-discipline team is a necessity. Two types of capabilities are required for an effective HAZOP review: (1) detailed technical knowledge of the process (chemistry, unit operations, controls, and automation) and (2) experience applying highly structured systematic HAZOP techniques.

The team leader (or chairperson or facilitator) of the HAZOP team must be selected for his or her ability to effectively lead the review and should have sufficient status to present the review recommendations to the proper level of authority. Ideally, the leader must be independent of the project.

For proper recording of the review discussion points, a secretary should be part of the HAZOP review team. He or she should be technically qualified to understand the review discussion and understand the HAZOP technique. The secretary should be preferably from a process discipline and understand the jargon used by the team.

Other team members for the HAZOP review must be selected on the basis of the positive contributions they can make based on their special knowledge and abilities. The participation of an operations and/or commissioning (launching) specialist for a similar plant or unit is strongly recommended. A typical team composition for a HAZOP review can be as follows:

### Full-Time or Core Team

- Leader
- Secretary
- Lead process engineer (preferably from both contractor or consultant and client)
- Control and automation engineer (preferably from both contractor or consultant and client)
- Operations and commissioning (launching) team, including process and certified and accredited engineers
- Project engineer (preferably from both contractor or consultant and client)

### Part-Time Team (Contractors or Consultants Engaged as Needed)

- Piping engineer
- Mechanical (static or rotary) engineer
- Electrical engineer
- Civil engineer

### Preparation

The HAZOP review info pack is a required component of the preparation for the review discussion. Additionally, the participants in the review have the duty to study the info pack in detail before the meeting. The team leader may suggest changes to the info pack to improve the team's understanding of the scope of the review.

Part of the preparation involves organizing the location, date, start time, and duration of the HAZOP review. Preparation is important and must start early enough to allow the team members to read the material and be ready to discuss. It is imperative that the info pack and meeting schedule be distributed to both full- and part-time members so that they can arrange their schedules. The responsibility for planning and dissemination of this information is with the project engineer.

The meeting location for the review must be spacious and well ventilated to enable the participants to be comfortable for long sessions. The meeting schedule must include periodic breaks with refreshments because of the intense nature of the discussions that require tremendous concentration. To make a review more effective, the meeting room must have:

- Wall projection equipment hooked to a computer. The equipment should be able to display the drawings and documents to be reviewed. The HAZOP review worksheet should be available as an editable document and be displayed on the projection wall for editing.
- Flip charts for recording new ideas or identifying issues for further investigation.
- Felt markers for making notes on drawings.
- Pins or tape for attaching drawings to walls.
- Laser light for pointing to certain areas of projected documents.

### Methodology

The entire process, plant, or unit is subdivided into manageable sections (nodes) for ease of understanding (see Figure I.3 in the Introduction to this book). Indications of nodes on P&IDs and their descriptions should be provided during the preparation stage and included in the info pack to save time during the review discussion.

The next step involves using a fixed set of terms (guidewords) for each process parameter (flow, temperature, pressure, etc.) in the selected node to identify a potential hazard or operability problem. The combination of guidewords and process parameters should reveal deviations in a process. For example, the NO guideword combined with the FLOW process parameter reveals a NO-FLOW deviation. Table 5.5 shows typical guidewords and parameters that may be used in a review process. The following steps are recommended for conducting a systematic HAZOP review:

- List node numbers on the HAZOP worksheet and use forms such as Table 5.3 and Table 5.5 or a format specific to your project to list the following details and steps for the node:
  - A brief description of the node in the HAZOP worksheet.
  - Process data such as operating pressure, operating temperature, design pressure, and design temperature for the node.
  - P&ID tag number covering the selected node. If more than one P&ID is required, provide all tag numbers.
  - Enter the first parameter in the appropriate column, for example, start with FLOW.
  - Enter the guideword against the parameter in the next column, for example, start with NO.
  - Identify the deviation. The deviation based on the two above items now is "No Flow."
  - Identify one cause for "No Flow" based on the process as depicted in the P&ID.
  - Outline the consequences (both upstream and downstream).
  - Identify the protection or safeguards available.
  - Provide the recommendations/actions to eliminate/mitigate the deviation.
  - Identify a second cause for "No Flow" and repeat the steps after the cause identification as above until the team agrees that no more reasonable causes of "No Flow" can be identified.
  - Repeat the entire procedure until all the deviations associated with the identified node have been considered.
  - After completion of an identified and numbered node move to another node and repeat the entire procedure as described above.
- Team members will apply their knowledge and experience to each deviation to identify possible causes (within scope of the HAZOP study) to establish the credibility of the event. The associated consequences will be considered to determine whether they significantly impact the hazards and operability of the plant or unit. Usually, there will be more than one cause for each deviation and the consequences may vary among causes.

## Recommendations

The results of a HAZOP review meeting should yield (1) suggested actions to prevent or mitigate deviations; (2) requests for additional information; and (3) recommendations for a quantitative risk assessment (QRA).

Typically, the results from the study will be produced in the form of an action list (corrective action report or CAR) and generate an individual action response form for each action point noted in the review meeting. The responsible individuals and required completion dates for each action item will be noted. The list will be subjected to periodic reviews to assess progress until the action items have been implemented and a HAZOP close-out report issued. In summary, all recommendations are intended to eliminate hazards or minimize them.

---

## Success Factors

All process improvement techniques involve fundamentals to keep in mind to achieve success. HAZOP is no different. In fact, the literature contains many recommendations and approaches, for example, Kletz (1999) and others. Crawley et al. (2000) devised a list of suggestions. We present them here not as a definitive approach but rather as guidelines. We chose their approach because it is simple and breaks down the effectiveness components in the order in which a HAZOP is conducted.

HAZOP also presents a number of pitfalls that must be addressed and eliminated before, throughout, and after the process. Listed below are common pitfalls that may affect the quality and value of a study.

### Before Study

A study must be initiated by a person who has authority to implement the results. If the person has no authority and the actions are not implemented, a study wastes time and money. The design analyzed must be well developed and firm; the operations or facilities examined should not be under development. A study cannot be carried out on a partly developed design because subsequent changes will undermine the results. Drawings must be accurate and depict what was studied. A study is worthless if the drawings are inaccurate or incomplete. Delay should be minimized because viable options will decrease. A skilled and suitably experienced team leader should be chosen. He or she must be given a clear scope, objectives, and terms of reference by the initiator, including delivery date and report distribution. If this is not done, a study may be incomplete and fail to fulfill the requirements of the initiator.

No study should be expected to yield project decisions and the design team should not adopt the approach of letting the HAZOP study decide what should be done. The study group must be balanced and well chosen to combine knowledge and experience. A study group drawn entirely from the project team will not be capable of a critical and creative design review. Equally, a group that has no operations background may lack objectivity. The group must be given adequate notice of the study so that they can carry out their preparations, including review and analysis of the P&IDs. The extent to which problems are evaluated, ranked, and solved should be defined.

## Throughout Study

Perhaps the most important factor in the success of a HAZOP is that it must form part of an overall safety management system. Another vital factor is the unequivocal support of senior management. The following issues are also important:

- The team must be motivated and have adequate time and resources to complete the examination.
- The boundaries of the study must be clearly analyzed. Changes of one item may impact other items involving different processes or an operation upstream or downstream. If the potential impact is not perceived correctly, the boundaries may be incorrect.
- The boundaries of a study of a modification are equally complex. A change in a reactor temperature may affect the by-product spectrum, thus producing a greater impact than the immediate modification. A clear description, design intention, and design envelope must be assigned to every section or stage examined.
- The study requires creative thought processes. If a study becomes mechanistic process or fatigue sets in, the study must be halted and restarted when the team is refreshed.
- Each action must be relevant, clearly defined, and described without ambiguity.
- If the person assigned to follow up the action has missed a meeting, the results of a misunderstanding could be wasted time and effort.
- The study must accept a flexible approach to actions. Not all actions are centered on hardware changes; procedural changes may be more effective.
- The study team members must be aware that some problems ranked and identified during the study may be caused by human factors and may not require hardware changes.
- Potential pitfalls must be treated individually when planning routes around branched systems such as recycling lines, junctions, vents, and drains.

## After Study

Every planned action must be analyzed and described accurately. Many of the actions raised will require no further change, but all must be designated for action or no action. Those that require changes should be subjected to a management of change process (that may require a HAZOP of the change) and put on a tracking schedule.

## Revisions

A risk management plan (RMP) is intended to *describe, communicate,* and *document* activities and processes necessary to manage the risks involved in the planned operations through all project phases. All processes and activities deemed necessary to manage risks during the operations should be reflected in the plan. The RMP should define and allocate responsibilities and serve as a tool for monitoring the status of the risk management process. The document should be established early, define responsibilities, and be maintained continuously to reflect the project status at various stages. Obviously, the complexity of a project will dictate the volume and detail of the information and discussion about low- and high-risk events. Some of the activities required for producing the RMP maybe controlled though a checklist such as:

- Establish health, safety, and environmental policies and strategies.
- Establish acceptance criteria.
- Define objects and operations for each item or process.
- Categorize potential risks for all items or operations.
- Define required risk identification and risk reducing activities.
- Establish follow-up and close-out routines.

It is imperative that the RMP is responsible for the specific activities cited on the checklist. All responsibilities should fall to management or another predefined responsible party. If several contractors or subcontractors are involved, the plan may be structured hierarchically, with each contractor executing a contract for his area of responsibility. Contractor and subcontractor RMPs should be based on plans and systems already in place. A first revision of the RMP is recommended as soon as the need for a change is realized. Further revisions should reflect the various project phases and stages. Table 5.7 lists revisions and recommendations.

**TABLE 5.7**

Revisions and Recommendations

| Point of Revision | Purpose of Issue |
| --- | --- |
| Project definition | Define responsibilities for risk management process and initial activities. |
| Completion of overall assessment | Communicate potential risk categories for planned operations after risks are identified and reducing activities are defined and/or when a contractor is nominated. |
| | Define details of all required risk identification and risk reducing activities, allocate responsibilities, schedule activities, and monitor status. |
| Project completion | Document completion dates of planned activities. |
| | Number of revisions and versions should reflect complexity and criticality of planned operations and number of contractors. |

# References

Crawley, F., M. Preston and B. Tyler. (2000). *HAZOP Guide to the Best Practice*. Warwickshire. Chemical Industries Association.

Det Norske Veritas. (2003). Risk management in marine and subsea operations. Høvik, Norway. http://www.srcf.ucam.org/polarauvguide/auvs/caseStudies/DNV-RP-H101a.pdf

HIPAP:8 (2008). *HAZOP Guidelines*. Sydney. New South Wales Department of Planning. Hazardous Industry Planning Advisory Paper 8.

Kletz, T. (1999). *Hazop and Hazan*, 4th ed. Manchester. Institution for Chemical Engineers.

Lihou, M. http://www.lihoutech.com/hzp1frm.htm

McGraw Hill (2003). *Dictionary of Scientific and Technical Terms*. 6th ed. New York: Author.

Pitblado, R.M., L. Bellamy, and T. Geyer (1989). Safety assessment of computer-controlled process plants. EFCE International Symposium on Loss Prevention and Safety Promotion in the Process Industries, Oslo.

# Selected Bibliography

American Institute of Chemical Engineers. (1994). *Guidelines for Preventing Human Error in Process Safety*. New York: Center for Chemical Process Safety.

American Institute of Chemical Engineers. (1985). *Guidelines for Hazard Evaluation*. New York: Center for Chemical Process Safety.

Andow, P. (1991). *Guidance on HAZOP Procedures for Computer-Controlled Plants*. London: U.K. Health and Safety Executive Contract Research Report 26.

Barton, J. and R. Rogers. (1997). *Chemical Reaction Hazards*, 2nd ed. London: IChemE.

Bullock, B.C. (1974). *The Development and Application of Quantitative Risk Criteria for Chemical Processes*. Fifth Chemical Process Hazard Symposium, Manchester, UK.

Elmendorf, M. (1996). Introduction to process hazard analysis. *Journal of Environmental Law and Practice*, 4, 36–56.

Geronsin, R. (2001). Job hazard assessment: a comprehensive approach. *Professional Safety*, 46, 23–30.

Gibson, S. B. (1974). Reliability engineering applied to the safety of new projects. *Chemical Engineering*, 306, 105.

http://paulthorn.co.uk/healthandsafety/Risk%20Management/HAZOP%20Guide%20to%20best%20practice-%20.pdf

Kletz, T. (1972). Specifying and designing protective systems. *Loss Prevention*, 6, 15.

Kletz, T. (1986). *Hazop and Hazan*. London: IChemE.

Kletz, T. (1995). *Computer Control and Human Error*. London: IChemE.

Kletz, T. (1998). *Process Plants: Handbook for Inherently Safer Design*, 2nd ed. London: Taylor & Francis.

Knowlton, R. (1981). *Introduction to Hazard and Operability Studies: A Guideword Approach*. Vancouver: Chemetics International

Knowlton, R. (1992), *Manual of Hazard and Operability Studies.* Vancouver: Chemetics International.

Lake, I. (1990) *HAZOP of Computer Based Systems.* Sydney: ICI Engineering.

Lees, F. (1996). *Loss Prevention in the Process Industries,* 2nd ed., Vols. 1–3. London: Butterworth-Heinemann.

Palmer, J. (2004). Evaluating and assessing process hazard analyses. *Journal of Hazardous Materials,* 115, 181–192.

Pitblado, R. and R. Tumey. (1996). *Risk Assessment in the Process Industries,* 2nd ed. London: IChemE.

Skelton, B. (1997). *Process Safety Analysis: An Introduction.* London: IChemE.

Standards Australia. AS 61508: Functional safety of electrical/electronic/programmable electronic safety-related systems.

# 6

## Fault Tree Analysis (FTA)

### Overview

Fault tree analysis (FTA) is another technique of reliability and safety analysis and one of many symbolic analytical logic methods applied in operations and system reliability research. Of course, other techniques such as reliability block diagrams (RBDs) shown later in this chapter, but FTA is more convenient and easier to use to evaluate safety and reliability issues.

Bell Telephone Laboratories developed the FTA concept in 1961 for used by the U.S. Air Force with its Minuteman system. FTA was later adopted and utilized extensively by the Boeing Company and is now used widely in many fields and industries.

FTA is a deductive analytical technique generally used for reliability and safety analyses of complex dynamic systems. It provides an objective basis for analysis and justification for changes and additions (Blanchard 1986). Fault tree diagrams (negative analytical trees) are logic block diagrams that display the state of a system (top event) in terms of its components (basic events); see Figure 6.1.

Like RBDs, fault tree diagrams are graphic design techniques that provide alternatives to RBD methods. A fault tree is built from the top down and utilizes events rather than blocks. It reveals a graphic model of the pathways within a system that can lead to foreseeable, undesirable loss events or failures. The pathways interconnect contributory events and conditions and utilize standard logic (AND, OR) symbols. The basic constructs of a fault tree diagram are gates and events; the events have the same meanings as blocks in RBDs; the gates represent conditions (Stamatis 2003).

As used today, the FTA model logically and graphically represents various combinations of faulty and normal events in a system that may lead to the top undesired event. It uses a tree to show the cause-and-effect relationships of a single, undesired event or failure and various contributing causes. The tree shows the logical branches from the single failure at the top to the root causes at the bottom (Figure 6.2) that may be analyzed further with an FMEA. Standard logic symbols shown in Table 6.1 and 6.2 are used. After the tree has been constructed and the root causes identified, the corrective

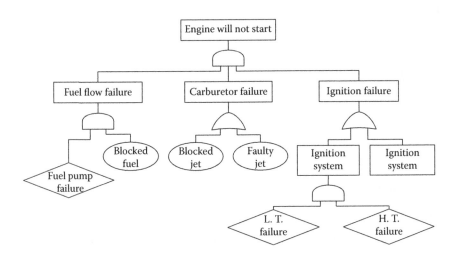

**FIGURE 6.1**
Typical partial engine FTA diagram.

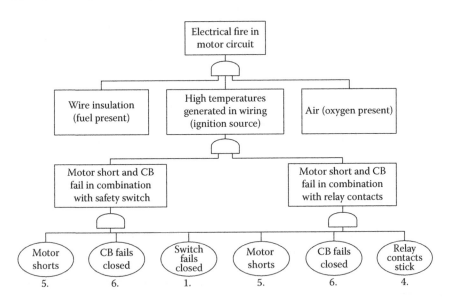

**FIGURE 6.2**
Relationship of FTA and FMEA.

actions required to prevent or control the causes can be determined. Usually probabilities are associated with undesired failures.

The FTA always supplements FMEA and not the other way around. In general, FTA may be applied to a system or subsystem environment with a focus on identifying the root factors and their interdependent relationships that could cause failures.

**TABLE 6.1**

Typical FTA Symbols

| Symbol of event | Meaning of symbol |
|---|---|
| Circle | Basic event |
| Diamond | Undeveloped event |
| Oval | Conditional event |
| House | Trigger event |
| Rectangle | Resultant event |
| In / Out Traiangle | Transfer-in and transfer-out events |

**TABLE 6.2**

FTA Logic Symbols

| Name of gate | Symbol of gate | Input-Output relationship |
|---|---|---|
| And gate | Output / Input 1 2....... $n$ | The output event occurs if all of the $n$ input events occur. |
| Or gate | Output / Input 1 2....... $n$ | The output event occurs if at least one of the $n$ input events occur. |
| $m$ out of $n$ voting gate | Output / Input 1 2....... $n$ | The output event occurs if $m$ or more of $n$ input events occur. |
| Priority and gate | Output / Input 1 2....... $n$ | The output event occurs if all input events occur in a certain order. |
| Exclusive or gate | Output / Input | The output event occurs if only one of the input events occur. |
| Inhibit gate | Output / Conditional event / Input | The input event causes the output event only if the conditional event occurs. |

**Benefits**

- Helps depict an analysis.
- Helps identify the reliabilities of higher-level assemblies or systems.
- Determines the probability of occurrence for each root cause.
- Provides documented evidence of compliance with safety requirements.
- Assesses the impacts of design changes and alternatives.
- Provides options for qualitative and quantitative system reliability analyses.
- Allows analysts to concentrate on system failure at a time.
- Provides insights into system behavior.
- Isolates critical safety failures.
- Identifies ways that a failure can lead to an accident.

## General Construction Rules

When constructing a fault tree, the scope of the analysis may need to be reduced to make it more manageable. This can be accomplished by using a block diagram such as the RBD for the system and equipment. A separate fault tree can then be constructed for each block of the RBD. Conversely, a success tree could be constructed for each block to identify the events required for the block to be a success.

After the top-level fault event has been defined, a series of steps should be followed by the FMEA team to properly analyze and construct the tree. A typical FTA is shown in Figure 6.1. The steps for constructing the FTA are:

1. Define the system and assumptions to be used in the analysis. Also, define what constitutes a failure (limit, parametric shift, functionality, etc.).
2. If needed to simplify the scope of the analysis, develop a simple block diagram of the system showing inputs, outputs, and interfaces.
3. Identify and list the top-level fault events to be analyzed. If required, develop a separate fault tree for each top level event, depending upon how the event is defined and the specificity of the event or scope of study.
4. Using the fault tree symbols and a logic tree format, identify all the contributing causes of the top-level event. In other words, using deductive reasoning, identify what events may cause the top-level event to occur.

5. Considering the causes revealed in Step 4 as intermediate effects, continue the logic tree construction by identifying causes for the intermediate effects.

6. Develop the fault tree to the lowest level of detail needed for the analysis, typically using basic or undeveloped events.

7. Analyze the completed tree to understand the logic and interrelations of the various fault paths and gain insight into the unique modes of product faults. Additionally, this analysis process should focus on faults that potentially appear most likely.

8. Determine where corrective action is dictated or a design change is required to eliminate fault paths or identify controls that prevent faults.

9. Document the analysis process and follow up to ensure that all appropriate actions have been taken.

After the appropriate logic gates, symbols, and event descriptions have been developed, the next level of complexity of FTA involves the calculation of the probabilities of occurrence of the top-level events. To perform the calculation, the probabilities of occurrence values for the lowest-level events are required. The probability equation is:

$$P_{(system\ fault)} = 1 - ([1 - P(1)]\ [1 - P(2)]\ [1 - P(3)])$$

where $P$ represents the probability of event occurrence. After the probabilities of occurrence of the lowest-level events have been determined, the probabilities of occurrence of the top-level events can be calculated by using Boolean algebra probability techniques. In most cases, it is more convenient to use computer software to perform these calculations; one is the *FaultrEase* program. An alternate approach to calculating the probability of occurrence of the top-level event is to convert the fault tree to its equivalent RBD.

To convert a fault tree, an OR gate corresponds to a series TBD and an AND gate corresponds to a parallel RBD. Recall that the reliability equation for a system is $R_{sys} = R_1 \times R_2 \times R_3$, where $R_1$ = reliability of Element 1 = $1 - P_{(Element\ 1\ Fault)}$ and $R_{sys}$ = system reliability = $1 - P_{(system\ fault)}$. The equation for $P_{(system\ fault)}$ is $P(1) \times P(2) \times P(3)$, where $P$ = probability of event occurrence. To convert to a parallel RBD, recall that the reliability equation is $R_{sys} = 1 - [(1 - R_1)\ (1 - R_2)\ (1 - R_3)]$, where $R_1$ = reliability of Element 1 = $1 - P_{(Element\ 1\ Fault)}$ and $R_{sys}$ = system reliability = $1 - P_{(system\ fault)}$. Figure 6.3 demonstrates these principles:

Using the formula for reliability a parallel system, $R_{sys} = 1 - [(1 - R_3)(1 - R_1 \times R_2)]$, where $R_1$ and $R_2$ represent reliabilities of Elements 1 and 2 in series. Therefore,

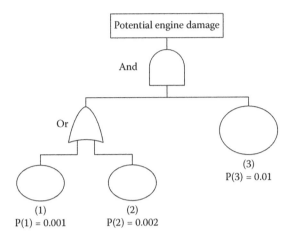

**FIGURE 6.3**
FTA depiction of parallel system.

$$R_{sys} = 1 - [(1 - R_3)(1 - R_1 \times R_2)]$$
$$= 1 - [(1 - 0.99)\{1 - (0.999)(0.998)\}]$$
$$= 1 - [(0.01)(1 - 0.997002)]$$
$$= 1 - [(0.01)(0.002998)]$$
$$= 1 - 0.00002998$$
$$= 0.99997002$$

Probability of damage $P(D) = 1 - R_{sys} = 0.00002998$ or approximately 0.0003. We must remind readers that the most fundamental difference between FTDs and RBDs is that one works in "success spaces" in an RBD and thus looks at system success combinations. A fault tree involves working in "failure spaces" and analyzes system failure combinations. Traditionally, fault trees have been used to determine fixed probabilities (each event on a tree has a fixed probability of occurring). RBDs may include time-varying distributions for success (reliability equation) and other properties such as repair and restoration distributions. Therefore, using the RBD (Figure 6.4) to analyze system reliability, we find the same answer as follows.

$R_1 = 1 - 0/001 = 0.999$; $R_2 = 1 - 0.002 = 0.998$; $R_3 = 1 - 0.01 = 0.099$. Therefore, $Rsys = 0.99997$ and $P(D) =$ probability of failure $= 1 - Rsys = 0.00003$. By using the RBD, the system reliability is $Rsys = 0.99997$. The probability of potential engine damage due to insufficient oil pressure $P(D)$ can be found by subtracting $Rsys$ from 1: $P(D) = 1 - Rsys = 1 - 0.99997 = 0.00003$. This, of course, agrees with the values of $P(D)$ obtained from the fault tree probability equations in Figure 6.3. This type of analysis enables both the product development, hazard team, and the FMEA team to determine the:

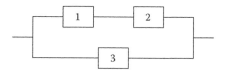

**FIGURE 6.4**
Typical block diagram.

Overall likelihood of the undesired top-level event

Combination of input events most likely to lead to the undesired top-level event

Single event contributing most to this combination

Most likely path(s) leading through the fault tree to the top-level event

Note that redundancy or system reconfiguration would improve the system reliability.

A final point for the FTA is a description of success trees. They complement fault trees and help identify contributors to robust functions. The FTA identifies possible causes of failure. Conversely, if we wish to design an ideal function, we may represent relationships by a success tree. A success tree is the complement or dual of a fault tree and focuses on what must happen for the top-level event to be a success. On the other hand, an ideal function is the mathematical representation of the inputs transformed 100% into expected or intended outputs.

Therefore, a fault tree can be converted to a success tree by changing each OR gate to an AND gate and changing each AND gate to an OR gate. Obviously, appropriately rewording statements within the blocks to their respective complements (success statements instead of fault statements) must be made. For a successful conversion, the same methodology and appropriate logic symbols used for a fault tree are used to construct a success tree.

This success tree–RBD relationship follows from the fact that for system success, the series RBD requires that Block 1 AND Block 2 AND Block 3 are all successes. Thus, the success tree AND gate corresponds to a series RBD and the success tree OR gate corresponds to a parallel RBD.

The benefit of a success tree is that it helps in the quantitative evaluation of a reliability prediction by using the same procedure used for fault trees. The top-level probability of occurrence of a success tree is probability of success, which by definition is reliability $R$. For this reason, a success tree can be used as a reliability prediction model.

After a success tree is developed for a product and the probability of occurrence determined for each cause, the reliability of the product can be determined by calculating the top-level probability of success.

## References

Blanchard, B. (1986). *Logistics Engineering and Management,* 3rd ed. Englewood Cliffs, NJ: Prentice Hall.

Stamatis, D. (2003). *Failure Mode and Effect Analysis: FMEA from Theory to Execution.* Milwaukee, WI: Quality Press.

## Selected Bibliography

Henley, E. and H. Kumamoto. (1981). *Reliability Engineering and Risk Assessment.* New York: Prentice Hall.

http://www.fault-tree.net/papers/clemens-event-tree.pdf

Kececioglu, D. (199 1). *Reliability Engineering Handbook,* Vols. 1–2. Englewood Cliffs, NJ: Prentice Hall.

Motorola Corporation. (1992). *Reliability and Quality Handbook.* Phoenix, AZ: Motorola Semiconductor Products Sector.

Omdahl, T.P., Ed. (1988). *Reliability, Availability, and Maintainability Dictionary.* Milwaukee, WI: Quality Press.

Stamatis, D.H. (2003). *Six Sigma and Beyond: Design of Experiments.* Boca Raton, FL: St. Lucie Press.

# 7

## Other Risk and HAZOP Analysis Methodologies

We already noted that risks and HAZOP issues may be analyzed in a number of ways, depending on the industry and scope of a project. This chapter focuses on available methods that are often overlooked for many reasons, including doubt about their effectiveness because of their simplicity.

### Process Flowchart

Flowcharts are easy-to-understand diagrams showing how steps in a process fit together. This makes them useful tools for communicating how processes work, and clearly documenting how a specific job is done. Furthermore, the act of mapping a process in a flowchart format helps clarify the understanding of the process and helps reveal where the process can be improved. A flowchart can therefore be used in HAZOP to:

- Define and analyze processes.
- Build a step-by-step picture of a process for analysis, discussion, or communication.
- Define, standardize, or find areas for improvement in a process.

Also, by conveying a process of operation in a step-by-step flow, one can then concentrate more intently on each individual step, without feeling overwhelmed by the bigger picture. To facilitate understanding, the method utilizes standard visual symbols for mapping the process flow. The most often used symbols are:

| | |
|---|---|
| Activity or operations | O |
| Inspection | □ |
| Flow or movement | → |
| Delay | D |
| Inventory storage | ∇ |

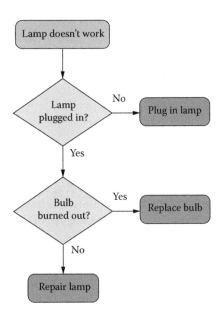

**FIGURE 7.1**
Simple flowchart.

A process flowchart is used primarily in HAZOPs, process FMEAs, and service FMEAs. A typical flowchart is shown in Figure 7.1.

## Functional Flow or Block Diagram

Block diagrams illustrate physical or functional relationships and interfaces within a system or assembly under analysis. They are used as a mechanism to depict system design requirements and illustrate operation series, parallel relationships, hierarchies of system functions, and functional interfaces. Specifically, a flow or block diagram is a pictorial representation of the reliability of a process. It is used to indicate the functioning components required for a process to perform and is applied primarily to systems in which the order of failure does not matter; see Figure 7.2. The types of block diagrams used in risk analysis are:

System: for identifying the relationships between major components and subsystems

Detail: for identifying the relationships between parts of an assembly or subsystem

Reliability: for identifying series dependence or independence of major components, subsystems, or detail parts in performing required functions

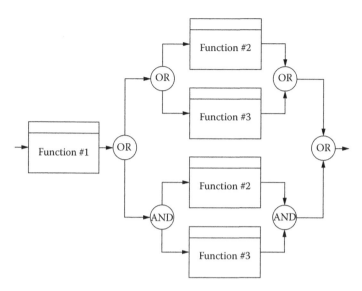

**FIGURE 7.2**
Logic depiction used in functional diagrams.

Block diagrams are not intended to illustrate all the functional relationships that must be considered in a HAZOP or FMEA. The diagrams should be as simple and explicit as possible. Figure 7.3 is a block and logic diagram.

System-level diagrams are generated for components or large systems comprising several assemblies or subsystems. Detail-level diagrams are generated to define the logical flow and interrelationships of individual components and/or tasks. Reliability-level diagrams generally are used at the system level to illustrate the dependence or independence of the systems or components contributing to specific functions (General Motors 1988). They also are used to support predictions of successful functioning for specified operating or usage periods.

Generally, it is assumed that a component has only two possible states: operational or faulty. To successfully apply the technique, the operation of a process must be described in detail. The description should contain statements of (1) functions to be performed; (2) performance parameters and possible limits; and (3) environmental and operating conditions.

A process is then divided into blocks that can be further divided into separate reliability block diagrams if required. If possible, each block should be independent of the other blocks and contain no redundancies. The system definition is then used to organize the block diagram. The output of one block is used as the input to the next. If no redundancy is present, the resulting reliability block diagram will be linear, indicating that the failure of any block will cause the entire process to fail. If redundancies exist within a process, blocks can be drawn in parallel, indicating that despite the failure of one of these blocks, a path for the process to work is still available.

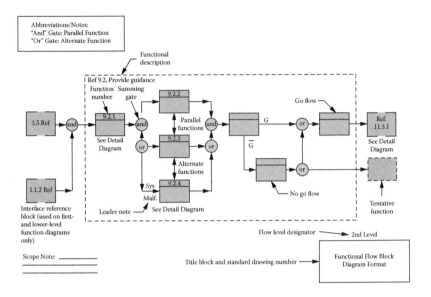

**FIGURE 7.3**
Block diagram with designated boundary line.

## Advantages and Disadvantages

- Advantages
  - Often used as a starting point for other techniques; can be used to identify areas where reliability is of concern and should be evaluated further by a more detailed technique.
  - Can identify where redundancy is required; specifically identifies where in-built redundancy will aid in safety.
- Disadvantages
  - Trivial except for complex systems; of little use on simple systems that perform only a few functions.
  - Limited to investigating reliability; cannot be easily applied to identify hazards not associated with reliability.

## Sketches, Layouts, and Schematics

These diagrams depict how a product or process is proposed to look (General Motors 1988). They are used to provide an analysis team with a better understanding of a system under study. The information in such diagram allows a team to gain objective information about:

- The relative sizes of components and the process operation
- Space requirements; how a component fits within a total system where about accessibility and serviceability are concerns
- Part and tool quantities; numbers of bolts, nuts, fixtures, tools, and other components of an operation

Stamatis (2003 pp. 55–58) provides examples of block diagrams, logic diagrams, schematic diagrams, functional diagrams, and layout diagrams.

## Failure Mode Analysis (FMA)

FMA is a systematic approach to quantify the failure modes, failure rate, and root causes of known failures. Usually, FMA is based on historical information such as warranty, service, field, and process data (Omdahl 1988). In a sense, FMA is a diagnostic tool because it concerns only known and/or occurred failures.

FMA is used to identify operations, failure modes and rates, and critical design parameters of existing hardware or processes. Because it utilizes historical data and known failures, FMA is used to analyze current production unlike the FMEA technique applied to modified and/or new designs, processes, and/or services. Both FMA and FMEA deal with failure modes and causes. FMA usually is done first, however, and results used to generate the FMEA.

## Control Plan

A control plan is a written summary of quality planning actions for a specific process, product, and/or service. The plan lists all process parameters and design characteristics considered important to customer satisfaction and requiring specific quality planning actions (Chrysler 1986; Ford 1992, 2000; General Motors 1988; ISO/TS 19649; AIAG 2001). A control plan describes the actions and reactions required to ensure that a process is maintained in a state of statistical control agreed upon by company and supplier.

Remember that FMEA identifies critical characteristics and therefore serves as the starting point for a control plan. A control plan cannot trigger the FMEA. A process flow diagram dictates the flow of the process. Stamatis (2003 pp. 60–61) provides an example of a control plan. A typical control plan may include:

- A listing of critical and significant characteristics
- Sample sizes and frequency of evaluation
- Method of evaluation
- Reaction and/or corrective action

## Process Potential Study (PPS)

The PPS is the experimental phase of new product introduction. It consists of implementing the control plan developed through a defined process and evaluates the results through statistical capability studies. If the desired capability is not achieved, changes in the control plan are defined, and the study repeated after the changes are made. This process may include designing experiments to determine optimum process parameters.

## Need and Feasibility Analysis

A feasibility analysis often is conducted as part of, or an extension to, a preliminary market analysis to:

- Define system operational requirements
- Develop a system maintenance concept
- Identify a system configuration that is feasible within the constraints of available technology and resources

The feasibility analysis uses product design and process FMEA as its primary tools (Ford 1992, 2000).

## Task Analysis

After a system has been defined and described, the specific tasks that must be performed are analyzed. Kirwan (1992) defines task analysis as a systematic method for analyzing a task based on its goals, operations, and plans. A task is a set of operations or actions required to achieve a set goal. The goal represents the required outcome of the actions, the operation involves various stages required to implement the goal, and plans are methods and conditions under which the stages are performed. In essence, a task analysis defines:

- Stimulus initiating the task
- Equipment used to perform the task
- Task feedback
- Required human response
- Characteristics of task output including performance requirements

Task analysis also studies the human activities involved in performing tasks and asks questions such as:

- What actions do the operators perform?
- How do operators respond to different cues in the environment?
- What errors may be made and what deviations of plant operations will result?
- How can errors be prevented or corrected and how can deviations be controlled?
- How do operators plan their actions?

A task analysis requires certain data:

- General operating procedure data, including job descriptions, process diagrams, and operating manuals
- Results of a hazard review
- Plant records
- Information obtained from interviews with people who understand the process and plant
- Observations of plant operations augmented by standards, exposure limits, and records of past incidents

The analysis is performed in a number of stages:

- *Setting goals of analysis:* The operation to be investigated should be fully described to the satisfaction of the assessment team and the requirements of the study defined.
- *Breakdown of operation into steps:* The original operation should be broken down to a level appropriate to the detail of the study—usually to the level of individual steps required to perform the operation; nonessential information may be removed.

- *Creating plan:* The methods and conditions under which the various stages of the analysis are performed should be defined.
- *Analyzing plan.* The plan created should be fully analyzed to identify hazards to the equipment, operators, and environment and should note lacks of controls and protection measures. Possible deviations and their likelihoods should also be examined.
- *Modifying plan:* Modifications to the plan should improve work methods and safety and minimize deviations. The plan should also recommend possible actions if deviations occur. The plan along with a description of the appropriate methods and conditions can be presented to management or other authority.

### Advantages and Disadvantages

- Advantages
  - Allows complex tasks to be analyzed in detail; splits complex tasks into simplistic components to allow detailed examination; components are easily understood and followed.
- Disadvantages
  - Only applicable to human interactions with process; is applied to tasks performed by plant operators and maintenance workers; technique cannot be easily applied to other areas.
  - Time consuming and expensive; a large number of tasks must be performed for a complex analysis and each needs to be fully developed to provide detailed information to the individual who must perform the task.

## Human Reliability Analysis

Swain (1983), Dougherty (1988), and Dhillon (1986) reported that HRA may be used to quantify human errors. The assessment is performed in a number of stages:

1. Define the system failures of interest.
2. List and analyze the related human operations.
3. Estimate the relevant error probabilities.
4. Estimate the effects of human errors on the system failure rate.
5. Recommend changes to the system and recalculate the system failure probabilities.

The first two stages are performed during task analysis and the results can be incorporated into the final stages of the HRA. The accuracy of the values produced by HRA is unknown and the results should be regarded only as estimates.

### Advantages and Disadvantages

- Advantages
    - Allows complex tasks to be analyzed in detail; incorporates task analysis and thus allows a detailed assessment of complex tasks.
    - Quantitative technique allowing limited predictions of human error. Probabilities are assigned to the human errors identified. The data can be used to determine the probability of human error over the entire task.
- Disadvantages
    - Only applicable to human interactions with a process; is tailored to assess human errors in performing tasks; cannot be easily applied to other areas of the process.
    - Time consuming and expensive. The technique evaluates large numbers of tasks and human errors; each error must be allocated a probability.
    - Relies on availability of human failure rate data for the lowest-level individual task.
    - Probabilities assigned to errors are often estimated; may produce errors in the calculation of the probability of an overall error.
    - Requires skilled practitioners or team members who must calculate realistic probabilities for errors and split tasks into their components.

---

## Failure Mode and Critical Analysis

FMCA is a systematic approach to quantify failure modes, rates, and root causes from a criticality perspective. It is similar to the FMEA in all other respects (Bass 1986). FMCA is used primarily with government contracts based on the MIL-STD-1629A, where the identification of critical, major, and minor characteristics is important.

One of the most important contributions of the FMCA is that focusing on criticality allows identification of the so-called single point failure modes. A single point failure mode is a single human error or hardware failure that can result in an accident (Motorola 1992).

## Hazard Identification (HAZID)

HAZID is a high-level systematic assessment of a plant, system, or operation intended to identify potential hazards. It is often used as a basis for risk assessment. Specifically, HAZID provides early identification of potential hazards and threats and the results serve as inputs to project development decisions. The benefit is a safer and more cost-effective design with fewer chances of design changes and cost penalties. Sometimes called "coarse HAZOP," HAZID uses a similar guideword approach but covers a wider scope of activities. HAZOP is concerned with deviations of process equipment; HAZID predominantly addresses the hazards outside the envelope of the process equipment (see Table 7.1). If a HAZID is conducted at an early stage, available formal documentation will be minimal or will show only preliminary details. Most of the information will be in the minds of the team. The following data should be made available:

- Process flow diagrams
- Substance information sheets
- Feasibility studies
- Survey reports
- Plant layouts
- Heat and mass balance data
- Project implementation plan
- Project description
- Standards and regulations
- Safety, design, and operation philosophies

A HAZID analysis involves six phases, each of which is distinct and requires specific tasks to be performed.

### Phase 1: Planning

- State the objectives of the risk assessment.
- Describe the activity to be evaluated.
- Confirm the scope (what will be included and excluded).
- Select team to perform the evaluation and assign responsibilities.
- Nominate a team leader.
- Identify how the results of the assessment will be communicated and to whom.

**TABLE 7.1**

Typical Hazards Outside Envelope of Process Equipment

| Parameter | Guideword | Parameter | Guideword |
|---|---|---|---|
| Hydrocarbon hazards | Layout | Natural environment | Extreme weather |
| | Over- and under-temperature | | Seismic activity |
| | Loss of containment | Transport | Vessels |
| | Flammable materials | | Road vehicles |
| | Fire protection | | Chemicals |
| | Overpressure | | Noise |
| | Different composition | | Toxics |
| | Ignition sources | Health hazards | Exposure |
| | Gas detection | | Working conditions |
| | Safety controls | | Radiation |
| Equipment or plant failure | Integrity | | Chemicals |
| | Material dissimilarities | | Noise |
| | Installation | | Toxic materials |
| | Failure modes | Damage to environment | Discharges to air |
| Utility systems | Capacity | | Pollution control |
| | Failure | | Discharges to sea |
| Operation and control | Different modes | | Waste management |
| | Normal shutdown | HSE management | Command and control |
| | Commissioning and start-up | | Emergency response |
| | Emergency shut-down | | Interfaces |
| Maintenance | Preparation | | Training |
| | Reinstatement | | Communications |
| | Execution | | Security |
| | Related tasks | | Staffing |
| | | | Supervision |

## Phase 2: Identifying Hazards

Three common methods are used to identify hazards:

- **Checklist:** The items listed depend on the industry and the process to be examined. Note that no checklist is ever all-inclusive. The team should be prepared to add or delete items as applicable to the specific study. Examples of checklist items are:
  - Electrical hazards
  - Environmental concerns
  - Human factors
  - Mechanical hazards
  - Process hazards
  - Quality issues
  - Radiation

- Thermal (fire) hazards
- Vibration
- Noise
- **HAZOP study:** If a HAZOP study is available for the equipment or process under study, use the report to generate a list of hazards. Be aware, however, that changes in operation practices and regulations may have occurred since the last study.
- **Brainstorming session:** Generate ideas in a meeting of experienced employees. These ideas generally are recorded on a risk assessment form in the "describe the hazard" column.

### Phase 3: Evaluating Hazards

Select hazards that fall within the scope of the assessment and have significant consequences. Describe what can go wrong and how. Generally, these details are recorded on a hazard and risk assessment form under "what can happen." Items to consider are:

- **People:** Injuries or occupational diseases affecting employees, customers, or the public
- **Environmental impacts:** Off-site or on-site contamination
- **Property damage:** Building fires, equipment damage, transport incidents
- **Business interruption:** Production outage, lost market share
- **Quality impacts:** Poor product quality or yield, customer dissatisfaction
- **Corporate image impact:** Negative image

### Phase 4: Assessing Risks

Select hazards that are likely to lead to failure and lead to consequences. Record the details of a hazard ("how it can happen") on a hazard and risk assessment. It must be noted here that human error is often the designated failure. When that happens the team should try to devise mistake-proof alternatives so that the human element of the failure is eliminated or at least minimized. In both cases, an error is an issue involving design and allowable risk. Some common types of potential failures are:

- Human errors of omission, such as an operator's failure to check equipment as required
- Human errors of commission, such as operator's inadvertent activation of a system

- Active equipment failures, such as pump stoppages or valves failing to close
- Passive equipment failures, such as pipe ruptures or structural failures

## Phase 5: Managing Risks

Managing risks requires answering two fundamental questions.

- **What controls are in place?** Identifying existing safeguards and determining their adequacy is imperative to the risk assessment process. Safeguards that do not fully prevent or mitigate hazards to an acceptable level cannot be considered true safeguards. When existing safeguards do not prevent or mitigate a hazard to an acceptable level, alternative safeguards listed in the "potential safeguards" section of the hazard and risk assessment form must be discussed and evaluated. The team should choose the safeguards that will best control the risks of a hazard and list them in the "adopted safeguards" section of the form.
- **How safe is it?** Considering the new or additional safeguards, the team should use a risk assessment matrix to determine the frequency, consequence or severity, and appropriate risk category for each item listed. If a new or additional safeguard has reduced a risk to an acceptable level, move on to the next item. If a risk is not adequately reduced, consider additional safeguards. After a risk category is assigned to each item, the next issue is to identify the risk control measures. The measures fall into design and administrative categories. Below is a list of generic risk control measures:
  - Eliminate hazard.
  - Substitute less hazardous product or situation.
  - Reduce hazard magnitude, volume, intensity, or complexity as required.
  - Enclose or isolate hazard.
  - Revise task procedures.
  - Institute training.
  - Require permits to work.
  - Install signs, tags, and lockout devices.
  - Modify maintenance systems to include preventative maintenance and condition monitoring.
  - Devise health and hygiene programs.
  - Utilize personal protective equipment (PPE).

## Phase 6: Monitoring Risks

Ensure that methods are in place to monitor risks. Monitoring can involve regular inspections, readings, alarm tests, operating procedure reviews, company policies, safety audits, or other administrative and engineering methodologies to ensure risks are properly controlled. Record the team's recommendations for monitoring the risks, along with the frequency in the appropriate section of the hazard and risk assessment form.

Monitoring a risk means that a system is in place for ongoing monitoring of risks. Inspections and audits are the most common monitoring and control methods. The level of risk dictates the frequency of monitoring. A risk assessment should be reviewed regularly to conform that risks continue to be reduced or eliminated. A review should indicate whether the assessment remains valid and that no changes of procedure have altered the risk level. The frequency of review should be stated in a risk assessment.

## Crisis Intervention in Offshore Production (CRIOP)

This is a structured method for assessing offshore control rooms. The main focus is to uncover potential weaknesses in accident and incident responses. CRIOP assesses the interfaces of operators and technical systems inside a control room. The assessment consists of (1) a design assessment in the form of a checklist and (2) a scenario-based assessment intended to assess the adequacy of responses to critical situations.

## Hazard Analysis and Critical Control Points (HACCP)

This technique involves seven steps or principles:

1. Analyze hazards. Potential hazards and measures to control them are identified.
2. Identify critical control points in production.
3. Establish preventive measures with critical limits for each control point.
4. Establish procedures to monitor the critical control points.
5. Establish corrective actions to be taken when monitoring shows that a critical limit has not been met.
6. Establish procedures to verify that the system works properly.
7. Establish effective recordkeeping to document HACCP systems.

These steps are presented as principles applying to the food industry as an example:

**Principle 1: Conduct a hazard analysis**—Plans should identify food safety hazards and the preventive measures for controlling them. A food safety hazard is any biological, chemical, or physical property that may cause a food to be unsafe for human consumption.

**Principle 2: Identify critical control points**—A critical control point (CCP) is a point, step, or procedure in a food manufacturing process where control can be applied and prevent, eliminate, or reduce a food safety hazard to an acceptable level.

**Principle 3: Establish critical limits for each critical control point**—A critical limit is a maximum or minimum value to which a physical, biological, or chemical hazard must be controlled at a CCP to prevent, eliminate, or reduce the hazard to an acceptable level.

**Principle 4: Establish critical control point monitoring requirements**—Monitoring activities are necessary to ensure that all processes are under control at all CCPs. In the United States, The Food Safety and Inspection Service (FSIS) requires that each monitoring procedure and its frequency be listed in HACCP plans.

**Principle 5: Establish corrective actions**—These actions must be taken when monitoring indicates a deviation from an established critical limit. The regulation requires a plant's HACCP plan to identify the corrective actions to be taken if a critical limit is not met. Corrective actions are intended to ensure that no product injurious to health or otherwise adulterated as a result of a deviation enters commerce.

**Principle 6: Establish procedures to ensure the HACCP system works as intended**—Validation ensures that plants perform as designed, that is, they successfully ensure production of a safe product. Plants are required to validate their own HACCP plans. FSIS will not approve HACCP plans in advance, but will review them for conformance with regulations.

Verification, on the other hand, ensures that HACCP plans are adequate—they work as intended. Verification procedures may include reviews of HACCP plans, CCP records, critical limits, and microbial sampling and analysis data. FSIS requires that HACCP plans include verification tasks to be performed by plant personnel. Verification tasks are also performed by FSIS inspectors. Both FSIS and industry undertake microbial testing as one of several verification activities. It is worth mentioning that verification often includes validation—the process of finding evidence for the accuracy of the HACCP system, for example scientific evidence backing up critical limitations.

**Principle 7: Establish recordkeeping procedures**—The HACCP regulation requires that all plants maintain certain documents including hazard analysis records, written HACCP plans, and monitoring data covering CCPs, critical limits, verification activities, and handling of processing deviations.

These seven principles are closely related to FMEA in the sense that they try to predict potential hazards. The difference is that FMEA focuses on the severity of a failure, then on occurrence, and finally on detection. HACCP focuses on hazards at critical points and then on controls. The techniques differ widely. From the author's view, FMEA is more powerful than HACCP.

In conjunction with HACCP, a reliability centered maintenance or remote condition monitoring (RCM2) evaluation is often performed. Again, FMEA is used to predict failures at the design and/or process level and devise action plans to avoid failures at those levels. RCM2 is a cost-effective life cycle asset management strategy. They are related but focus on different areas. FMEA evaluates failure issues; RCM2 is concerned with costs. One may use FMEA data in CRM2 but the reverse will not work.

## Near-Miss Reporting

Although near-miss reporting is not exactly a methodology, we include it because it is an important tool for recognizing potential failures of many kinds. Adequate near-miss reporting, handling, and registration are important steps for reducing the numbers of incidents and accidents to an absolute minimum and obtaining feedback. Near-miss reporting does not conflict with the zero-incident mind-set. It is a management tool intended to ensure that all personnel understand on the importance of reporting near-misses and are encouraged to report them.

The application of all near-misses during operations must be reported according to defined procedures by all personnel who witness an incident, preferably on a standardized reporting form. Completed forms must be distributed to specified (appropriate) recipients.

All personnel are responsible for reporting near-misses to the relevant management. Simple near-miss reporting forms should be available throughout operations areas, usually in clearly marked boxes designed for this purpose. All near-misses must be listed in a special log and reviewed in the daily operations meetings.

Recording near-misses is not enough. Corrective actions must be taken to avoid repetitions. Near-miss reporting and corrective action procedures must be components of project documentation.

## Incident and Accident Investigation and Reporting

Another way of handling failures, safety issues, and hazards is incident and accident reporting. The need for investigation after reporting must be assessed on a case-by-case basis in accordance with procedures set by the organization. The project management group must instigate investigations of all serious accidents. A dedicated accident investigation task force must be named and given a mandate for the investigation. A task force leader must be appointed. He or she must have knowledge of the operation and have experience in investigating events that have safety ramifications. All task force personnel must be independent of the project.

The task force will normally investigate an incident or accident by interviewing all relevant personnel and reviewing all relevant project documentation. The task force must conclude its work by issuing an incident or accident investigation report and presenting it to the project manager or other appropriate management in draft form. All relevant comments concerning misunderstandings and errors must be considered and corrected by the task force before the report is issued in final form.

The purpose of reporting is to prevent recurrences and also inform management and relevant authorities required by laws and regulations. Due to legal ramifications, all incidents and accidents surrounding an operation must be reported appropriately.

Anyone who witnesses an incident or accident must report it as soon as possible to the appropriate management level and legal authority if necessary. Management must make sure the situation is stabilized and contained. In case of serious incidents or accidents, all relevant witnesses must describe their observations in writing in appropriate on prescribed forms. Reports of incidents or accidents must contain the following information:

- Location of incident or accident
- Time of incident or accident
- Detailed description
- Descriptions of injuries, if any, including names of injured Individuals
- Description of damages
- Immediate actions taken
- Names of reporting personnel

Every incident or accident must be reported appropriately and discussed in the daily meeting of the operation where the event happened. Long-term corrective actions must be identified and implemented if needed to prevent recurrence.

## Semi-Quantitative Risk Assessment (SQRA)

The intent of SQRA is to subjectively assess the risk and criticality of operations to identify the most critical activities. The technique assesses risks for each undesired event identified and reported in a hazard review (PHA, HAZID and/or HAZOP). One concern is that the risks and/or criticalities are subjectively assessed by an evaluation team of qualified persons. Table 7.2 is a typical work sheet for HAZID and SQRA.

The deliverables are based on an assessment of relative risks. For that reason, it is imperative to appoint team members who are both experienced with and knowledgeable about risks, failures, and hazards that affect the operation, task, or project at hand. Ranking of the most critical activities and operations that may later serve as a basis for project risk registers are generally prioritized based on a top-ten ranking or similar method.

## Audits

An effective audit includes a review of the relevant documentation and process safety data, inspection of the facilities, and interviews with all levels of plant personnel. By using an audit procedure and checklist developed in the preplanning stage, an audit team can systematically analyze compliance with standards and any other relevant corporate policies. For example, the audit team may review all aspects of a training program as part of its assignment. It will review written materials for adequacy of content, frequency of training, and effectiveness of training in meeting goals and objectives fitting the relevant standards.

Through interviews, the team can determine employees' knowledge and awareness of safety procedures, duties, rules, and emergency response assignments. During the inspection, the team can observe work practices, including safety and health precautions. This approach enables the team to identify deficiencies and determine where corrective actions or improvements are necessary.

## Event Tree Analysis (ETA)

ETA is based on binary logic: an event has or has not happened or a component has or has not failed. The technique is valuable for analyzing the consequences arising from a failure or undesired event. ETA is generally applicable

**TABLE 7.2**

Worksheet for HAZID with SQRA

| Operation or Activity | Undesired Event | Description of Consequences | Existing Risk Reducing Measures (Probabilities and/or Consequences) | Risk Value $P \times C = R$ (SQRA only) | Actions or Measures to Reduce or Eliminate Risk | Responsibility for Implementing Actions | Comments |
|---|---|---|---|---|---|---|---|
| | | | | | | | |
| | | | | | | | |
| | | | | | | | |
| | | | | | | | |

$P$ = probability. $C$ = consequence. $R$ = risk value.

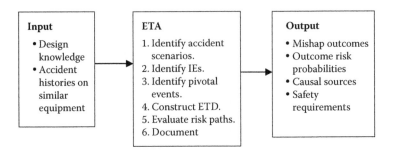

**FIGURE 7.4**
Overview of ETA.

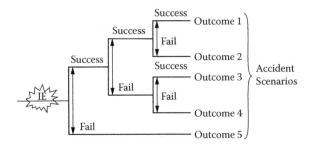

**FIGURE 7.5**
Generic ETA showing primary and secondary trees.

for most risk assessment applications but is most effective for modeling accidents in which multiple safeguards are in place as protective features. ETA is highly effective in determining how various initiating events can cause accidents. A visual overview is shown in Figure 7.4. ETA gives an analyst the ability to handle large-scale problems and utilize success logic. The event tree model may be created independently of the fault tree model or may use fault tree analysis gate results as sources of event tree probabilities.

It is important to note that ETA can handle both primary and secondary event trees, multiple branches, and multiple consequence categories; see Figure 7.5. Some of the possibilities of ETA are:

- Primary and secondary event trees
- Multiple branching support for event trees
- Multiple consequence categories for event trees
- Pruning of event trees
- Inserting new columns while retaining existing data
- Copying and pasting

As flexible as the ETA is, it has other unique features worth mentioning. They are based on the ease of application such as identifying:

- Full minimal cut set analysis allowing full handling of success states
- Sensitivity analysis allowing the automatic variation of event failure and repair data within specified limits
- Range of event failure and repair models, including fixed rate, dormant, sequential, stand-by, time-at-risk, binomial, Poisson, and initiator failure models
- Risk importance analysis identifying major contributors to risk
- Basic events that may be linked to Markov analysis
- Comprehensive risk calculation

An event tree begins with an initiating event and progresses through component failures and concludes with outcomes. The flow is shown in Table 7.3. The consequences of an event may follow a series of possible paths. Each path is assigned a probability of occurrence and the probabilities of various possible outcomes can be calculated. A typical flow for a fire example is shown in Table 7.4. As the arrows in the cells indicate, a concern is identified in the

**TABLE 7.3**

Flow of ETA

| Initiating Event | Critical Events | | | Outcomes |
|---|---|---|---|---|
| | Event 1 | Event 2 | Event 3 | |
| | | | | |
| | | | | |
| | | | | |
| | | | | |

**TABLE 7.4**

Flow of ETA in Application Format

| Fire Start | Fire Detected | Fire Alarm Start | Sprinkler System Start | Consequence | Results |
|---|---|---|---|---|---|
| | | | | | ⟶ |
| | | | | | |
| | | | | | |
| | | | | | |
| | | | | | |

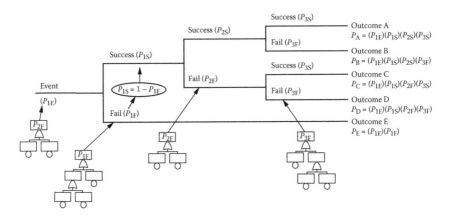

**FIGURE 7.6**
Generic ETA associated with FTA and propabilities.

left-most cell. The team progressively follows the details of the event until the results are identified in the extreme right cell. Generally the flow follows a binary functional diagram. Figure 7.6 shows probabilities associated with ETA and FTA. In developing the ETA, one may want to use FTA and associated probabilities for each event identified. A typical generic format may look like Figure 7.6.

## Characteristics

ETA models a range of possible accidents resulting from an initiating event or category of initiating events. A typical ETA is shown in Figure 7.7. In essence, ETA is a risk assessment technique based on success and failure that effectively accounts for timing, dependence, and domino effects among various accident contributors that are too cumbersome to model in fault trees. The assessment is performed primarily by an individual working with subject matter experts through interviews and field inspections with the intent to generate at least:

- Qualitative descriptions of potential problems as combinations of events producing various types of problems (ranges of outcomes) from initiating events
- Quantitative estimates of event frequencies or likelihoods and relative importances of various failure sequences and contributing events
- Lists of recommendations for reducing risks
- Quantitative evaluations of recommendation effectiveness

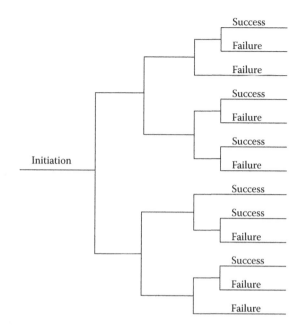

**FIGURE 7.7**
Typical ETA showing individual events of success and failure.

## Process

The ETA process follows seven steps:

1. **Define the system or area of interest.** Specify and clearly define the boundaries of the system or area for which ETA will be performed.

2. **Identify the initiating events of interest.** Conduct a screening level risk assessment to identify the events of interest or categories of events to be addressed. Categories include such events as vessel groundings, collisions, fires, explosions, toxic releases, and so on.

3. **Identify lines of assurance, physical phenomena, and safeguards (lines of assurance) that will help mitigate the consequences of the initiating event.** Lines of assurance include engineered systems and human actions. Physical phenomena include ignitions and meteorological conditions that may affect the outcome of an initiating event.

4. **Define accident scenarios.** For each initiating event, define the scenarios that can occur. Do not be afraid to identify events that may be considered "wild" or "out of the box."

5. **Analyze accident sequence outcomes.** Determine the appropriate frequency and consequences that characterize each specific outcome on the tree.

6. **Summarize results.** Event tree analysis can generate numerous accident sequences that must be evaluated. Summarizing the results in a separate table or chart will help organize the data for evaluation.

7. **Use the results in decision making.** Evaluate the recommendations from the analysis and the benefits they are intended to achieve. Benefits can include improved safety and environmental performance, cost savings, or additional production. Determine implementation criteria and plans. The ETA results may serve as bases for deciding whether to perform additional analyses on a selected subset of accident scenarios.

Keep in mind that ETA is used to determine the path from an initiating event to various consequences and reveal the expected frequency of each consequence. Pipe breaks, alarms that fail to activate, and human errors of omission and commission are events that can produce insignificant or catastrophic consequences. The **event tree** models these initiators and consequences and determines their frequencies. In summary, ETA is one more way to evaluate hazards. It is a bottom-up deductive analytical technique that is applicable to automated and human-operated systems and to decision-making and/or management systems.

Specifically, ETA can explore system responses to initiate challenges and opportunities for pursuing successes and assessing failures; see Figure 7.7. Furthermore, ETA is closely related to other techniques, especially FTA and FMEA; see Figure 7.8.

Typical challenges that may be analyzed using the ETA are ignitions of stored combustibles, epidemic outbreaks, utility system failures, technology needs, business competition, and others. ETA is simply a credible system of analyzing operating permutations that lead to a success or failure. This, of course, is the classic Bernoulli model. Binary branching will reveal unrecoverable failures and undefeatable successes leading to final outcomes. Of course, after the ETA is formulated, other analyses such as FTA may be necessary to determine the probability of an initiating event or condition; see Figures 7.6 through 7.8.

## Advantages and Disadvantages

- Advantages:
  - End events need not be foreseen.
  - Potential single point failures can be identified.
  - System weaknesses can be identified.

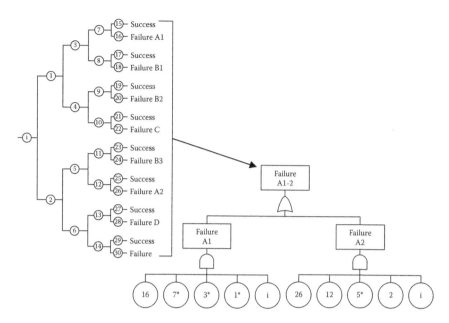

**FIGURE 7.8**
ETA and FTA relationship.

- Multiple failures can be analyzed.
- Zero payoff system elements or options can be discarded.
- Disadvantages:
  - Partial successes and or failures are not distinguishable.
  - Operating pathways must be anticipated.
  - Sequence-dependent scenarios are not modeled well.
  - Initiating events are treated singly. Multiple events require multiple trees. Coexisting events are not covered in an initial tree.

## Example

Clemens (1990) presents a simple example of an ETA dealing with an anti-flooding system (Figure 7.9). Figure 7.10 is a reliability diagram of the system and Figure 7.11 shows its reliability and associated probabilities. A subgrade compartment containing important control equipment is protected against flooding by the system shown. Rising floodwaters will close float switch S, powering pump P from an uninterruptible power supply. A klaxon (horn) K sounds to alert operators to perform manual bailing B should pump P fail. Pumping or bailing will dewater the compartment effectively. We will assume flooding has commenced and analyze responses of the dewatering system. The assumptions for this system are:

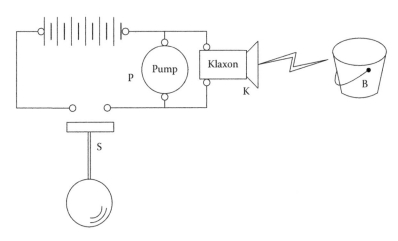

**FIGURE 7.9**
Anti-flooding system.

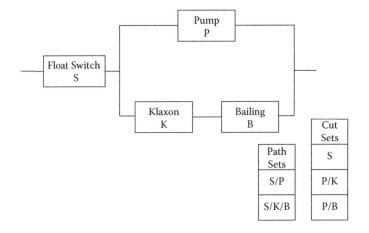

**FIGURE 7.10**
Reliability diagram of flooding system.

- Power is available full time.
- Only four system components (S, P, K, and B) are considered.
- Operator error as included within the bailing B function.

To conduct an ETA, it is necessary to:

- Develop an event tree representing system responses.
- Develop a reliability block diagram for the system.
- Develop a fault tree for the top event: failure to dewater.

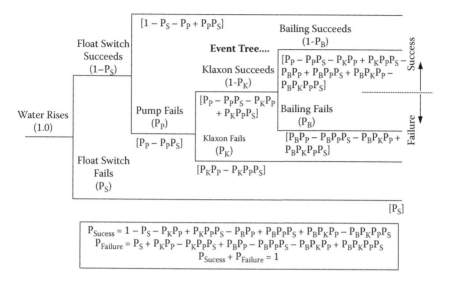

**FIGURE 7.11**
ETA reliability diagram with associated probabilities.

# References

AIAG, Ed. (2000) *ISO Technical Specification 19649: Quality Management Systems.* Daimler Chrysler Corporation, Ford Motor Company, and General Motors Corporation. Southfield, MI: Author.

AIAG. (2001). *Potential Failure Mode and Effect Analysis,* 3rd ed. Daimler Chrysler Corporation, Ford Motor Company, and General Motors Corporation. Southfield, MI: Author.

Bass, L. (1986). *Product Liability: Design and Manufacturing Defects.* Colorado Springs, CO: Shepard/McGraw Hill.

Chrysler Motors. (1986). *Design Feasibility and Reliability Assurance in FMEA.* Highland Park, MI: Author.

Clemens, P. (1990). *Event Tree Analysis,* 2nd ed. Sverdrup. http://www.fault-tree.net/papers/clemens-event-tree.pdf

Dhillon, B. (1986). *Human Reliability with Human Factors.* Oxford, U.K.: Pergamon Press.

Dougherty, E., Jr. and J. Fragola. (1988). *Human Reliability Analysis: A Systems Engineering Approach with Nuclear Power Plant Applications.* New York: John Wiley & Sons.

Ford Motor Company (1992). *FMEA Handbook.* Dearborn, MI: Author.

Ford Motor Company (2000). *FMEA Handbook with Robustness Linkages.* Dearborn, MI: Author.

General Motors Corporation (1988). *FMEA Reference Manual.* Detroit, MI: Author.

Kirwan, B. and L. Ainsworth, Eds. (1992). *A Guide to Task Analysis.* New York: Taylor & Francis.

Motorola Corporation (1992). *Reliability and Quality Handbook.* Phoenix, AZ: Author.

Omdahl, T. P., Ed. (1988). *Reliability, Availability, and Maintainability Dictionary.* Milwaukee, WI: Quality Press.

Stamatis, D.H. (2003). *Failure Mode and Effect Analysis (FMEA) from Theory to Execution*, 2nd ed. Milwaukee, WI: Quality Press.

Swain, A. and H. Guttman. (1983). *Handbook of Human Reliability Analysis with Emphasis on Nuclear Plant Applications*. U.S. Nuclear Regulatory Commission, Report NUREG-CR-1278.

---

## Selected Bibliography

Andrews, J. and S. Dunnett. (2000). Event tree analysis using binary decision diagrams. *IEEE Transactions on Reliability*, 49, 230–238.

Center for Chemical Process Safety. (2008). *Guidelines for Hazard Evaluation Procedures*, 3rd ed. New York: John Wiley & Sons.

Ericson, C. (2005). *Hazard Analysis Techniques for System Safety*. New York: John Wiley & Sons.

Gibson, S. (1974). Reliability engineering applied to the safety of new projects. *Chemical Engineering*, 306, 105.

Gould, J., M. Glossop, and A. Ioannides. (2000). *Review of Hazard Identification Techniques*. Health and Safety Laboratory. HSL/2005/58

Henley, E. and H. Kumamoto. (1981). *Reliability Engineering and Risk Assessment*. New York: Prentice Hall.

Henley, E. and H. Kumamoto. (1996). *Probabilistic Risk Assessment and Management for Engineers and Scientists*, 2nd ed. New York: IEEE Press.

Kaplan, S. and B. Garrick. (1981). On the quantitative definition of risk. *Risk Analysis*, 1, 11–37.

Kletz, T. (1972). Specifying and designing protective system. *Loss Prevention*, 6. 15.

Lees, F. (2001). *Loss Prevention in the Process Industries*, 2nd ed., Vols. 1–3, Maryland Heights: MO: Elsevier-Butterworth-Heinemann.

Papazoglou, I. (1998). Functional block diagrams and automated construction of event trees. *Reliability Engineering and System Safety*, 61, 185–214.

Stamatis, D. H. (2003). *Six Sigma and Beyond: Design for Six Sigma*. Boca Raton, FL: St. Lucie Press.

Stamatis, D.H. (2003). *Six Sigma and Beyond: Design of Experiments*. Boca Raton, FL: St. Lucie Press.

# 8

## Teams and Team Mechanics

This chapter covers the basic aspects of teams and how team actions affect both HAZOP and FMEA results. The information in this chapter does not represent an exhaustive examination of teams but does cover issues related to both methodologies. To achieve best results, a risk analysis (HAZOP, FMEA, FTA, etc.) must be written by a team. This is because a risk analysis should act as a catalyst to stimulate interchanges of ideas among the groups affected (Stamatis 1991). A typical view is shown in Figure 8.1.

A single engineer or other individual cannot perform a risk analysis. A team should consist of five to nine people (preferably five). All team members must have some knowledge of team behavior, the task at hand, the problem to be discussed, and direct or indirect ownership of the problem. Above all, they must be willing to contribute. Team members must be cross-functional and represent varied disciplines. Furthermore, whenever possible, customers and/or suppliers should participate as ad hoc members.

### Team Members, Qualifications, and Activities

A team is a group of individuals committed to achieving common organizational objectives. They meet regularly to identify and solve problems and improve processes. They work and interact openly and effectively together and produce desired economic and motivational results for an organization. Figure 8.2 shows these relationships. Three factors and several subfactors influence the performance and productivity of a team:

1. Organization
   a. Philosophy
   b. Rewards
   c. Expectations
   d. Norms
2. Team
   a. Meeting management
   b. Roles and responsibility
   c. Conflict management

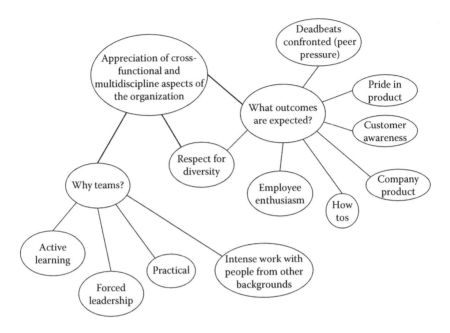

**FIGURE 8.1**
Team overview.

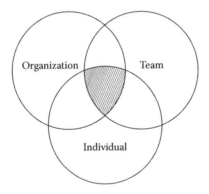

**FIGURE 8.2**
Team performance factors.

    d. Operating procedures

    e. Mission statement

  3. Individual members

    a. Self-awareness

    b. Appreciation of individual differences

    c. Ernpathy

    d. Caring

## Benefits of Using Teams

*Synergy* is the theory that encourages the team formation instead of individual effort. Synergy dictates that the sum of the total is greater than the sum of the individuals. Another description is that two heads working together are better than two heads working individually. From a risk analysis perspective, a team is the foundation of improvement. The team defines the issues and problems in a specific task environment, identifies and proposes ideas, recornmends appropriate measures, and provides decisions based on consensus. Generally speaking, a team is formed to address concerns about:

- Work
  - Task complexity
  - Productivity and quality advantages
  - Work system stability
- People
  - Rising expectations
  - Affiliation needs
  - Increased cognitive ability
  - Specific time-related concerns
  - Future directions
  - Survival in a global market

All teams regardless of application must be familiar with problem solving steps:

- Statement of the problem
- Root cause analysis
- Solution based on facts
- Implementation
- Evaluation

Another requirement is a clear charter issued by management to:

- Define the task
- Have accountability and responsibility
- Define the boundaries
- Define and communicate the barriers
- Have the authority to request support
- Have a clear authority to implement recommendations within the scope of the team

To paraphrase Allmenclinger (1990), to harness the collective intelligence (synergy) of a team to benefit an organization, the following must take place:

- **Relevancy:** The information gathered by the team should be of value.
- **Reliability:** The process for collecting information should be consistent and isolated as much as possible from staff and organizational changes.
- **Accuracy:** The data should be expressed in a manner that most accurately reflects its information content; accuracy should not be confused with precision.
- **Efficiency:** The design and implementation of the tasks should minimize the burden imposed by the data collection process.

## HAZOP Team

An effective hazard incident investigation and analysis program generally has two major components: technical and human. The technical side of the investigation is where most of the literature focuses, particularly on root cause analysis. However, the human aspects of incident investigation do not receive the same degree of attention. An effective investigator understands how people think and behave. He or she must be able to communicate with a wide range of team members and management levels. Chapter 10 covers effective communication processes.

### Technicians

Most hazard incidents involve front-line technicians (operators and maintenance workers), some of whom may have been injured or emotionally shaken. These people will often feel defensive and upset and may feel guilty if any colleagues were injured or died.

Technicians often may not understand what caused an incident and worry that they will be blamed. An effective investigator encourages these front-line technicians to be open and candid, primarily by letting them talk without interruption. Unfortunately, many investigators, even those with years of experience are quick to interrupt a technician's narrative flow with questions, war stories, or snap judgments about the event. An investigator should also clearly state that the goal of the investigation is to learn what happened, not apportion blame or show how smart the investigator is.

### Mid-Level Managers

Most investigations find that changes are needed at the facility's mid-level management systems. Examples of such changes include an increased emphasis on equipment inspection, upgraded operating procedures, and more training for the technicians. The implementation of such changes requires that the facility managers commit scarce resources that they would prefer to spend on achieving other goals. An effective investigator will empathize with these mid-level managers and will understand the demands that are being placed on the organization by the investigation and its follow-up.

### Senior Managers

Many investigators find that technicians are candid and open and mid-level managers are generally willing to honestly address the need for improvements to systems. What investigators sometimes learn, however, is that senior managers can be resistant to the findings and implications of an investigation. The findings may indicate that systemic changes to management systems are required. Senior managers in charge of such systems may become very defensive about holding onto their ideas. An effective investigator will know how to communicate with senior managers and obtain their buy-ins. This ability is critical because senior managers provide the funding needed to implement the investigation's recommendations.

An additional concern about the involvement of senior managers is that they are usually strong personalities; they may try to take over an investigation and direct it to meet their own opinions, goals, and agendas. A strong investigator is able to resist these efforts.

A HAZOP team will typically consist of five to nine people. Team members should be cross-functional and possess a range of relevant skills to ensure all aspects of the plant and its operations are covered. All relevant engineering disciplines, management, and plant operating staff should be represented. This will help prevent possible events from being overlooked through lack of expertise and awareness.

A fundamental difference between regular and HAZOP teams is that a regular team operates under the direction of a leader who may be selected by members and can facilitate meetings. A HAZOP team operates under the direction of a chairperson who must be knowledgeable about HAZOP techniques and has experience in conducting HAZOPs. This will ensure that the team follows the procedure without diverging or taking shortcuts. If a HAZOP is required as a condition of development consent, the name of the chairperson is usually submitted to the regulators and/or other authority that approves the commencement of the exercise.

---

## Consensus

Consensus is a collective decision reached through active participation by all members who have personal ownership in the decision. It requires all members to express their views, actively listen, and differ constructively. Consensus does not mean 100% agreement. It is a decision about which a team member can state reservations ("I am not totally sold on it," I am not 100% sure," "I do not agree completely") and still support or "live with it."

In a team environment, discussion continues until the team reaches a decision that every member accepts and will support even if some members have reservations. Ideally, the team capitalizes on diversity of members to reach a better decision than they could produce independently. Consensus decision making takes time, and like any other skill, requires practice to achieve proficiency.

A classic example of true consensus can be seen in the *Twelve Angry Men* movie. One juror (Henry Fonda) holds an opinion that is opposite to the opinions of the other 11 jurors. It is interesting to note that Fonda thought a guilty verdict was appropriate. His desire to discuss the case from a different perspective, however, made the case interesting and relevant to a discussion of team consensus. By the end of the movie, all 12 jurors made a decision based on probable doubt. Despite unanswered questions about the case, they all agreed that doubts persisted and they rendered an innocent verdict.

For a team to reach consensus, four requirements must be met:

1. The process requires 100% participation.
2. Members must actively participate, listen, and voice their disagreements in a constructive manner.
3. The requirement is for 100% commitment to support a decision, not 100% agreement to the decision.
4. Majority does not rule. Minority sometimes generates a correct decision or a single individual may be on the right track.

To reach consensus, the team members must:

- Be open to influence and new ideas.
- Contribute, not defend.
- Actively listen to other points of view.
- Learn the reasons for others' positions.
- Avoid averaging the differences.
- Confront the differences politely.
- Stand up for their own thoughts and opinions.

To recognize consensus, the team members must answer yes to four questions:

- Have I honestly listened?
- Have I been heard and understood?

- Will I support the decision?
- Will I say, "We decided," as opposed to "My idea went through," or "I decided," or "I told them and they followed my recommendation"?

---

## Team Process Check

For a team to be effective and productive, an occasional process check may be appropriate (Stamatis 1991). This is very important, especially for very complex problems the team may face. Some of the items covered in a process check are:

- Purpose of meeting not clear
- Meeting held only to "rubber stamp"
- Repetition of old information
- Too boring; trivial matters discussed
- Leader tendency to lecture team
- Team members not prepared
- Vague assignments
- No summary
- Lack of time or ability to deal with the unexpected

Errors will occur if a team continues to meet without a process check. Errors may be prevented by testing, training, and auditing. Some of the most common errors are:

- Results of misunderstandings
- Inadequate information
- Incomplete data because form is too difficult to complete
- Incomplete or biased data caused by fear
- Failure to use existing data

### Difficult Team Members

The idea of forming a team is to help members learn from each other and share knowledge. All members must participate in activities. That dynamic is not guaranteed because some individuals participate at the expense of others or do not participate at all. Problem individuals fall into three general classifications (Jones 1980; Stamatis 1987, 1992).

**Member who talks too much:** If a discussion turns into a dialogue between the leader and an overly talkative individual, the other members will lose interest. Even if the talkative individual has something of value to say, the team leader should not let him or her monopolize the discussion. Tactful approaches must be used to divert the discussion to others. If the leader knows that one member of the team likes to dominate, he or she should pose questions to the group without looking at the talkative individual or ignore the individual's responses.

The leader of a FMEA or HAZOP team should want all members to participate. If one member dominates the discussion because he or she has more experience or education, the leader should utilize that individual as a resource and a coach for other team members.

A talkative person may simply be trying to make an impression to satisfy his or her own ego. The only way to handle such an individual is to advise him or her in advance that the group disapproves of such behavior and continue exerting team pressure as needed.

A talkative person may interfere by taking too much time to express his or her ideas. This is a very sensitive area because the person participates as expected but also annoys the rest of the team. The leader must handle this situation delicately. If the situation is mishandled, the talkative participant will lose self-confidence and ultimately withdraw. Usually, it is better to tolerate a certain amount of this behavior rather than discourage the individual too much.

Another case is the talkative person who starts a private conversation with a neighbor. A leader can eliminate this problem by asking a direct question to those involved in the conversation or make the team large enough to allow generation of a variety of ideas and small enough to sustain small cliques. A team usually consists of five to nine persons; a five-person team is the most common.

**Member who talks too little:** Members may not want to participate, may feel out of place, or may not understand the problem discussed. It is the responsibility of the leader to actively draw this individual into the discussion by direct questions at meetings or by attempting to motivate the individual outside the team environment with statements such as "We need your input," "We value your contribution," or "You were selected for the team because of your experience and knowledge."

**Member who strays from agenda:** This problem is common in a team environment (especially early in team development) where individuals want to talk about their own agendas instead of the issues facing the team. It is the responsibility of the leader to bring the discussion back to the meeting agenda and/or outline. On rare occasions the leader may want to involve the whole team by asking whether the off-agenda item should be "taken up right now" or "sent to the 'parking lot' for future discussion."

## Problem Solving

This section will not discuss problem-solving tools and methods. It focuses on helping teams to understand the mechanics and rationale for pursuing methods to eliminate and/or reduce problems. Detailed descriptions of tools may be found in basic Statistical Process Control (SPC) books, statistical literature, and/or organizational development sources. For a good review of the process, readers may want to see Stamatis (2002, 2003).

For most people, the prediction or onset of a problem indicates a need for a change in behavior. When an individual or a team is actually or potentially in trouble, a unique set of strategies is required to trigger at least a temporary change in behavior—a new course of action. Without a deliberate strategy for pursuing a new course of action, revised behaviors may make a situation worse.

Problems are often not clear to the individuals who experience them. It is difficult to isolate a problem and its related components. Even if this is possible, the selection and implementation of a solution involves some physical or psychological risk. Familiar patterns of behavior are safe. In a problem situation, a person is torn between the need to change and the desire to maintain the old patterns. This conflict produces strong emotions and anxieties that affect the cognitive processes required to make workable decisions. If a problem is sufficiently severe, cognitive paralysis may result (Pfeiffer 1991).

People and teams who are in trouble need useful tools to help them understand their problem situations, decide their courses of action, and manage the new directions chosen. The components of a generic model of problem identification and solving are:

Stage 1: Identify the problem.

Stage 2: Determine scope: gather facts and organize data.

Stage 3: Define the problem.

Stage 4: Analyze the problem, list alternative solutions, and select a feasible one.

Stage 5: Implement the solution.

Stage 6: Evaluate the solution.

Stage 7: Follow-up by monitoring the solution.

Stage 8: Continually improve.

These steps constitute a summary of techniques presented over the years. Details of specific tools for each stage may be found in Stamatis (2003). Many of the tools ignore the drives, emotions, needs, preferences, values, and conflicts that surround most human problems. Furthermore, they are of little use in attacking problems that people consider intangible. The techniques may be useful for evaluating an alternative business plan or buying a new washing machine, but offer little help in interpersonal problems. Therefore, this section discusses methods of incorporating human issues into the problem-solving process.

## Meeting Planning

Before a team actively works on a project, some preliminary steps must be followed. The first step is planning the meeting. Bradford (1976); Nicoll (1981); Schindler-Rainman, Lippit, and Cole (1988); and Stamatis (1991) identified the following concerns.

**People:** Meeting participants may differ in values, attitude, experience, sex, age, and education. All these differences, however, must be considered in planning a meeting.

**Purpose:** The purpose, objective, and goal of the meeting must be understood by all participants and by management.

**Atmosphere:** The comfort of participants contributes to the effectiveness of a meeting. It is imperative that the meeting planner consider the climate and atmosphere.

**Place and space:** Planners must consider (1) access to meeting space; (2) size of space; (3) acoustics, lighting, and temperature control; (4) costs; (5) equipment required; and (6) available parking.

**Costs:** Cost is of paramount importance. The preparation of a risk analysis through PHA, HAZID, HAZOP, FMEA, FTA, etc., is a lengthy process. Another consideration is that the system, design, process, and service personnel involved in the project may be in different and distant places.

**Time considerations:** How long will this activity take? Is an alternate schedule available? Can the participants be spared for this task? Without evaluating time constraints and recognizing that a meeting may be prolonged for an unexpected reason, the agenda items and objectives may suffer.

**Pre- and post-meeting work:** The amount of work resulting from a meeting is related directly to the amount of planning that preceded the meeting. Lengthy and complex tasks may require major portions of work to be performed outside the meeting and only reviewed by participants at the meeting.

**Plans, program, and agenda:** An agenda is essential and no meeting can proceed without one. A detailed program or agenda distributed to all participants ensures effectiveness and prevents surprises. An agenda should cover all objectives of the meeting.

**Beginning, middle, and end:** Every meeting, regardless of duration and significance, has a beginning, middle, and end. Proper planning is essential. Without it, failure to focus on the agenda will cause an unproductive gathering and the group will fail to meet its objectives.

**Follow-up:** After a meeting ends, all agenda items should have been addressed and resolved. If some items remain open, a further meeting should be scheduled to address such items. Follow-up may be required to (1) implement action items, (2) communicate information to appropriate parties, (3) prepare and distribute minutes, and (4) write reports.

Management and the team leader should meet to identify the hazards in question. Basic potential hazard terms are:

- Struck by
- Struck against
- Caught between
- Contact with
- Contacted by
- Caught in
- Caught on
- Fall from same level
- Fall to lower level
- Overexertion
- Exposure

After basic hazards are identified, the team should consider the categories that the hazards in question belong. The major categories are:

**Chemical :** Effects on health, the flammability potential, and reactive capabilities

**Biological:** Exposure to viruses, fungi, mold, or bacteria

**Physical:** Temperature extremes, noise, electrical energy, or vibrations

**Psychosocial:** Environment, organizational culture, management style, content of task

**Mechanical:** Imposition of physical stress

**Human factors:** Impact caused by human error

Some specific typical hazards and their descriptions are:

Chemical and toxic effects on health

Flammability potential

Corrosive damages or material changes

Explosive chemical reaction

Over-pressurization causing sudden release

Exposure to electrical shock or short circuit

Fire caused by overheating or ignition of flammable material

Loss of power causing critical safety equipment failure

Ergonomic strain from overexertion or repetition of movements

Human error arising from inadequate or faulty procedure or equipment

Excavation collapse caused by improper shoring

Slip, trip, or fall

Burns and organ damage caused by exposure to fire or heat

Mechanical vibration that damages nerve endings

Material fatigue

Mechanical failure or operation of equipment beyond capacity

Skin, muscle, or other injury caused by mechanical malfunction

Noise levels exceeding 85 dBA (8-hour time weighted average [TWA])

Ionizing radiation exposure from x-rays, alpha, beta, and gamma rays

Non-ionizing radiation exposure from ultraviolet, infrared, visible light, and microwaves

Struck by falling or moving objects

Struck against a surface

Extreme heat or cold, heat stress, heat exhaustion, hypothermia

Lack of visibility, inadequate lighting, obstructed vision

Exposure to snow, rain, wind, ice

Other site-specific hazard considerations

At this stage, the team should have a good understanding of the hazard and have developed ideas about resolving it. The team has two options: (1) eliminate the risk or (2) reduce it. In both cases, the team has three choices of corrective actions:

1. Engineering approach: Eliminate, enclose, isolate, remove, or redirect the hazard.

2. Administrative approach: Develop written procedures, limit exposure times, monitor exposures to hazardous materials, install warning devices, and mandate training.

3. Use of personal protective equipment (PPE). This approach is acceptable only when other controls are not feasible.

---

## In-Process Meeting Management

In addition to more detailed meeting planning, managers may find it necessary to spend more time managing meeting participants. Every organization has an overt hierarchy involving job titles and designations of managers and subordinates. Every organization also has an unspoken hierarchy that involves a pecking order even among people who occupy the same rung of the hierarchical ladder. Some members dominate; they are talkative and tend to interrupt others. Less aggressive members may not feel comfortable challenging dominant members and may remain silent at meetings. The result is uneven participation that often leads to boredom and ineffective communication. Mosvick and Nelson (1987) identified 11 steps to ensure effective meeting management and foster decision making:

1. State and restate the initial question until everyone agrees on the issue to be discussed.
2. Solicit participants' honest opinions at the outset.
3. Think of opinions as hypotheses; test them instead of arguing over them.
4. Plan a method of testing opinions against reality by reviewing the issue and the goal.
5. Establish a rule that additional information revealed at a meeting must be relevant to agenda topics.
6. Encourage disagreements and differences of opinions.
7. Do not judge others' opinions hastily. Learn to appreciate diverse points of view.
8. Encourage members' commitments to resolving the issue whenever possible.
9. Compromise as needed.
10. Ask whether a decision is necessary. Remember that choosing to do nothing is a legitimate choice.
11. Construct a process for feedback to determine whether a decision was successful.

## Common Meeting Pitfalls

The following dysfunctional patterns and behaviors commonly appear at meetings (Bradford 1976):

- Vying for power, often by challenging the leader or by wooing a group of supporters, thus dividing the group.
- Excessive joking and clowning that distracts participants and may disguise hostility.

- Failing to agree on an issue or problem.
- Arguing about others' opinions or suggestions; this stifles the brainstorming process and can cause embarrassment or discomfort.
- Wandering off the topic at hand.
- Forcing meeting members to answer to the chairperson (usually someone higher on the organizational ladder).

Awareness of these traps can help a meeting facilitator avoid them. Constructive confrontation is an effective technique for dealing with many disruptive and dysfunctional meeting behaviors. A meeting leader who chooses to confront must discuss only behavior, not the participant. More desirable behaviors should be suggested in a direct but caring way. Jones (1980) suggests two approaches to dealing with disruptive meeting participants. The first requires the meeting leader to communicate directly with the disruptive person by:

- Turning his or her question into statements, thus forcing the person to take responsibility for his or her opinion.
- Refusing to engage in a debate. By noting that debates have winners and losers, the leader should promote a win–win outcome.
- Suggesting that the leader and disruptive person swap roles to demonstrate the effect of the disruptive person on the group.
- Using active listening techniques to mirror a participant's feelings, for example, "You seem upset today, especially when I disagree with you." We have two ears and only one mouth. Therefore, listen twice as hard as you speak.
- Agreeing with the person's need to be heard and supported.

The second approach suggested by Jones treats the other meeting participants as allies against the disruptive person:

- Ask the participants to establish norms that will discourage you're-right-I'm-wrong thinking.
- Post all participants' inputs anonymously on flip charts to make information available to all and eliminate repetition.
- Break the participants into small groups. This immediately limits a dominating person's sphere of influence. Give the groups assignments that require them to reach consensus.

## Utilizing Meeting Management Guidelines

Meeting leaders may find that small teams can help prevent the participants from falling into the common meeting traps. When people break into small teams for discussion, less assertive members often become more willing to participate. A small team is not as likely to wander off the subject as a large team. Because fewer people compete for attention in a small team, members feel a stronger sense of commitment. Finally, small teams can diffuse aggressive members' tendencies to dominate discussions.

Meeting leaders will find that their meetings will become more interesting, lively, and balanced if they follow the guidelines presented in this section. The core points to remember are that all meeting participants must be treated equally; honesty must be the norm; and all opinions must be respected (Stamatis 1992).

---

## References

Allmenclinger, G. (1990). Performance measurement: impact on competitive performance. *Technology,* December, 10–13.

Bradford, L.P. (1976). *Making Meetings Work: A Guide for Leaders and Group Members.* San Diego, CA: University Associates.

Jones, J. E. (1980). Dealing with disruptive individuals in meetings. In Pfeiffer, J. S. and E. Jones, Eds., *The 1980 Annual Handbook for Group Facilitators.* San Diego, CA: University Associates.

Mosvick, R. K., and R. B. Nelson. (1987). *We've Got to Start Meeting Like This! A Guide to Successful Business Meeting Management.* Glenview, IL: Scott, Foresman.

Nicoll, D. R. (1981). Meeting management. In Pfeiffer, J. S. and E. Jones, Eds., *The 1981 Annual Handbook for Group Facilitators.* San Diego, CA: University Associates.

Pfeiffer, J. W., Ed. (1991). *Theories and Models in Applied Behavioral Science: Management Leadership,* Vols. 2–3. San Diego, CA: University Associates.

Schindler-Rainman, E., R. Lippit, and J. Cole. (1988). *Taking Your Meetings out of the Doldrums.* San Diego, CA: University Associates.

Stamatis, D. H. (2003). *Six Sigma and Beyond: Statistical Process Control.* Boca Raton, FL: St. Lucie Press.

Stamatis, D. H. (2002). *Six Sigma and Beyond: Problem Solving and Basic Mathematics.* Boca Raton, FL: St. Lucie Press.

Stamatis, D. H. (1987). Conflict: you've got to accentuate the positive. *Personnel,* December, 47–50.

Stamatis, D. H. (1991). *Team Building Training Manual.* Southgate, MI: Contemporary Consultants.

Stamatis, D. H. (1992). *Leadership Training Manual.* Southgate, MI: Contemporary Consultants.

# 9

## OSHA Job Hazard Analysis

The U.S. Department of Labor's Occupational Safety and Health Administration (OSHA) issued Guideline 3071 (revised 2002) covering job hazard analysis. This chapter discusses some of the suggestions as they play a role in a HAZOP evaluation. The guidelines are available on the Internet (http://www.osha.gov/Publications/osha3071.pdf).

A hazard is the potential for harm, often associated with a condition or activity that, if left uncontrolled, can lead to injury or illness. In Chapter 8, we identified several common hazards and descriptions. Table 9.1 identifies specific categories of hazards. Identifying hazards and eliminating or controlling them as early as possible will help prevent injuries and illnesses.

A job hazard analysis (JHA) focuses on job tasks as a way to identify hazards before they occur. The technique is different in scope from a traditional HAZOP that focuses on processes. The JHA focuses on the relationship of worker, task, tools, and work environment. Ideally, after uncontrolled hazards are identified, an organization will take steps to eliminate or reduce them to acceptable risk levels.

As noted in previous chapters many workers are injured and killed every day around the world. Safety and health measures can add value to your business, your job, and your life. You can help prevent workplace injuries and illnesses by analyzing your workplace operations, establishing proper job procedures, and ensuring that all employees are trained properly. One of the best ways to determine and establish proper work procedures is to conduct a job hazard analysis as one component of a wider commitment to develop a complete safety and health management system.

The result of this analysis may help supervisors to find potential and existing job hazards and eliminate or prevent them. The analysis should result in fewer worker injuries and illnesses; safer, more effective work methods; reduced workers' compensation costs; and increased productivity. The analysis also can be a valuable tool for training new employees to perform their jobs safely. For a JHA to be effective, management must demonstrate its commitment to safety and health and correct any uncontrolled hazards identified by the analysis. Otherwise, management will lose credibility and employees may hesitate to consult management when dangerous conditions threaten them.

**TABLE 9.1**

Specific Hazards by Categories

| Type | Description |
|---|---|
| Electrical | Electrical hazard is any use of electrical power that results in electrical overheating or arcing to the point of combustion or ignition of flammables or electrical component damage. It may also involve moving or rubbing wool, nylon, other synthetic fibers, or flowing liquids that can generate static electricity. An excess or deficiency of electrons on a material surface discharges (sparks) to the ground, leading to the ignition of flammables, damage to electronics, and damage to the human nervous system. A typical electrical hazard is a critical equipment failure resulting from a power loss that causes ergonomic harm (strains and sprains) due to overexertion or repetitive motion. Ergonomics may affect and effect system design, process design, procedures, and equipment. The focus in ergonomic design is to eliminate and or minimize situations leading to human errors, for example, designing a switch that clearly indicates on and off conditions. |
| Mechanical | Mechanical hazards occur when devices fail because they operate beyond capacity or are inadequately maintained. Mechanical issues may affect skin, muscle, or other body parts exposed to impact, crushing, cutting, tearing, shearing and may be viewed as equipment failures. Other failures are noise levels exceeding 85 dBA (8-hour time weighted average) that lead to hearing damage or inability to communicate safety-critical information; ionizing radiation (alpha, beta, gamma, and neutral particles) and x-rays that cause damage of cellular components); non-ionizing radiation (visible, ultraviolet, infrared, and microwaves) that injure tissues by thermal or photochemical means. |
| | Other examples are soil collapses in trenches or excavations resulting from improper or inadequate shoring. Soil type is critical in determining hazard likelihood. Fall conditions (slippery floors, poor housekeeping, uneven surfaces) lead to slips and trips. Fire, heat, and extreme temperatures can cause burns and other organ damage. A fire requires a heat source, fuel, and oxygen. Vibration can damage nerve endings. Vibration or material fatigue can cause a safety-critical failure. Examples are abraded slings and ropes and weakened hoses and belts. |
| | Other mechanical hazards are falling objects and projectiles that can cause injury or death by striking a body part, for example, a screwdriver that slips; temperatures that result in heat stress, extreme exhaustion, or metabolic slow-down (hypothermia); poor visibility (inadequate lighting or obstructed vision) that results in an error or accident. Weather is an obvious mechanical hazard. |
| Chemical | A chemical hazard involves absorption of a toxic material through the skin, lungs, or bloodstream that causes illness or death. |
| | The amount of exposure is critical in determining hazardous effects. MSDSs and OSHA 1910.1000 provide chemical hazard data. |
| | Some chemicals combust when exposed to a heat ignition source. Typically, the lower a chemical's flash point and boiling point, the more flammable it is. Check MSDSs for flammability information. |
| | Corrosive chemicals such as acids and bases cause skin damage and may destroy surrounding materials. |
| | A chemical alone or a chemical reaction may also cause an explosion (a sudden and violent release of a large amount of gas and/or energy due to a significant pressure difference), for example, the rupture of a boiler or compressed gas cylinder. An explosion may be triggered by contact with an exposed electric conductor or a short circuit, for example, a metal ladder that touches power lines. The common domestic 60 Hz alternating current can stop a heart. |

A job hazard analysis can be conducted on many jobs but certain types of jobs have high priorities:

- Jobs with the highest injury or illness rates
- Jobs with the potential to cause severe or disabling injuries or illnesses, even if there is no history of previous accidents
- Jobs in which a simple human error could lead to a severe accident or injury
- Jobs that are new to your operation or have undergone changes
- Jobs complex enough to require written instructions.

Traditionally, there are five ways to evaluate a starting point for a JHA.

**Involve your employees:** It is very important to involve employees in the hazard analysis process. They have a unique understanding of the job and their knowledge is invaluable for finding hazards. Involving employees will help minimize oversights, ensure a quality analysis, and obtain worker buy-ins to solutions because they will share ownership in their safety and health program.

**Review your accident history:** Review with your employees the work site's history of accidents and occupational illnesses that needed treatment, losses that required repairs or replacements, and near-misses (incidents that did not cause accidents or losses but could have). These events are indicators that existing hazard controls may not be adequate and deserve more scrutiny.

**Conduct a preliminary job review:** Talk with your employees about the hazards in their work areas. Brainstorm with them for ideas to eliminate or control the hazards. If any hazards pose immediate dangers to an employee's life or health, take immediate action. Any problems that can be corrected easily must be corrected as soon as possible. Do not wait for completion of a job hazard analysis. This will demonstrate your commitment to safety and health and enable you to focus on the hazards and jobs that need more study because of their complexity. Evaluate types of controls for hazards determined to present unacceptable risks. Some typical hazard controls are shown in Table 9.2.

Information obtained from a job hazard analysis is useless unless hazard control measures recommended in the analysis are implemented. Managers should recognize that not all hazard controls are equal. Some are more effective than others at reducing risks. The order of precedence and effectiveness of hazard control is typically as outlined in Table 9.2.

The selection of one hazard control method over another higher in the control precedence may be appropriate for providing interim protection until a hazard is abated permanently. In reality, if a hazard cannot be eliminated entirely, the adopted control measures will likely be a combination of all three types of control measures instituted simultaneously.

**TABLE 9.2**

Hazard Categories and Controls

| Control Category | Examples |
| --- | --- |
| Engineering | Elimination or minimization of hazard by designing facility, equipment, or process to remove the hazard, or substituting processes, equipment, materials, or other factors to lessen it |
| | Enclosure of hazard (enclosed cabs, enclosures for noisy equipment) or other means |
| | Isolation of hazard with interlocks, machine guards, blast shields, welding curtains, or other means |
| | Removal or redirection of hazard, e.g., improved local and exhaust ventilation |
| Administrative | Written operating procedures, work permits, and safe work practices |
| | Limits on exposure times (usually to control temperature extremes and ergonomic hazards) |
| | Monitoring uses of highly hazardous materials |
| | Installation of alarms, signs, and warnings |
| | Implementation of buddy system |
| | Train employees to find and avoid hazards |
| Personal protective equipment (PPE) | Respirators, hearing protection, protective clothing, safety glasses, and hard hats; PPE is acceptable as a control method (1) if engineering controls are not feasible or do not totally eliminate a hazard; (2) while engineering controls are being developed; (3) if safe work practices do not provide sufficient protection; and (4) during emergencies when engineering controls may not be feasible |

**List, rank, and set priorities for hazardous jobs:** List jobs with hazards that present unacceptable risks starting with those most likely to occur and those presenting the most severe consequences. These jobs should be your first priority for analysis.

**Outline the steps or tasks:** Most jobs can be broken down into tasks or steps. When beginning a job hazard analysis, watch an employee perform and list each step he or she takes. Record enough information to describe each action without excessive detail. Avoid making the breakdown of steps so detailed that it becomes unnecessarily long or broad and does not describe basic steps. It may be valuable to obtain input from other workers who perform the same job. After the observation step, review the steps with the employee to make sure no steps are omitted. Inform the employee that you are evaluating the job and *not* his or her performance. Include the employee in all phases of the analysis from reviewing the job steps to discussing uncontrolled hazards and recommended solutions. In conducting a job hazard analysis, it may be helpful to photograph or videotape the worker performing the job. These visual records can be helpful in conducting a more detailed analysis of the work.

After the five steps have been addressed, job hazard analysis becomes an exercise in detective work to determine:

- What can go wrong
- The consequences
- How it can occur
- Other contributing factors
- Likelihood of occurrence

To make a JHA useful, document the answers to these questions in a consistent manner. Describing a hazard in this way helps ensure that the efforts to eliminate it and implement controls will target the most important contributors. Good hazard scenarios describe:

- Where it may happen (environment)
- Who or what will be affected (exposure)
- What precipitates the hazard (trigger)
- The outcome (consequence)
- Other contributing factors

Table 9.3 is a typical hazard analysis form that helps organize the pertinent information and relevant. Rarely does a hazard arise from a single cause and produce a single effect. Usually many factors line up in a certain way to create a hazard.

Here is a simple example of a hazard scenario in a metal shop (environment). While clearing a snag (trigger), a worker's hand comes into contact (exposure) with a rotating pulley that pulls his hand into the machine and

**TABLE 9.3**

Hazard Analysis Form

| Job title | |
|---|---|
| Job location | |
| Analysis date | |
| Task description | |
| Hazard description | |
| Consequences | |
| Hazard controls | |
| Rationale or additional comments | |

severs his fingers (consequences) quickly. In a job hazard analysis, the above items are considered:

- What can go wrong? The worker's hand may come into contact with a rotating object that "catches" the hand and pulls it into the machine.
- What are the consequences? The worker may be severely injured and lose hands and fingers.
- How could it happen? The accident resulted when the worker tried to clear a snag during operation or as a maintenance activity while the pulley rotated during operation. Obviously, this hazard scenario could not occur if the pulley was not rotating.
- What are other contributing factors? This hazard occurs quickly; the workers has no opportunity to recover or prevent injury once his hand comes into contact with the pulley. This is an important factor, because it helps determine the severity and likelihood of an accident and the selection of appropriate hazard controls. Unfortunately, experience has shown that training is not very effective in hazard control when triggering events happen quickly because humans can react only so quickly.
- How likely is it that the hazard will occur? This answer requires some judgment. If previous incidents or near-misses occurred, the likelihood of recurrence Is high. If the pulley is exposed and easily accessible, that raises another consideration. In the example, the likelihood that the hazard will occur is high because no mechanical guard prevents the contact of hand and pulley and the operation is performed while the machine is running.

The steps in this example allow us to organize hazard analysis activities. The next example shows how a JHA can identify existing or potential hazards for each basic step involved in grinding iron castings. The job involves three steps. Table 9.4 shows the analysis.

Step 1: Reach into metal box to right of machine, grasp casting, and carry to wheel.
Step 2: Push casting against wheel to grind off burr.
Step 3: Place finished casting in box to left of machine.

After reviewing the list of hazards with the employee, consider what control methods will eliminate or reduce them. The most effective controls are engineering modifications that change a machine or work environment to prevent employee exposure to the hazard—a mistake-proof approach. The more reliable or less likely a hazard control can be circumvented, the better. If this is not feasible, administrative controls may be appropriate, for example,

**TABLE 9.4**

Hazard Analysis of Grinding Castings

| Job Location | Metal Shop |
| --- | --- |
| **Name of Analyst** | Stacey Robinson |
| **Date** | 1/5/13 |

| | Step 1 | Step 2 | Step 3 |
| --- | --- | --- | --- |
| Task description | Worker reaches into metal box to right of machine, grasps 15-pound casting; carries it to grinding wheel; grinds 20 to 30 castings per hour | Worker reaches into metal box to right of machine, grasps 15-pound casting; carries it to grinding wheel; grinds 20 to 30 castings per hour | Worker reaches into metal box to right of machine, grasps 15-pound casting; carries it to grinding wheel; grinds 20 to 30 castings per hour |
| Hazard description | Worker could drop casting onto his foot; casting size and weight and height could seriously injure foot or toes | Castings have sharp burrs and edges that can cause severe lacerations | Reaching, twisting, and lifting 15-pound castings from floor level could strain muscles of lower back |
| Hazard controls | (1) Worker removes castings from box and places them on table next to grinder; (2) wears steel-toe shoes with arch protection; (3) wears protective gloves that allow better grip; (4) uses device to pick up castings | (1) Worker uses clamp or other device to pick up castings; (2) wears cut-resistant gloves that allow good grip and fit tightly to minimize the chance of being caught in grinding wheel | (1) Move castings from floor level and place them closer to work zone to minimize lifting; ideally place them at waist height or on adjustable platform or pallet; (2) train workers not to twist while lifting and reconfigure work stations to minimize twisting during lifts |

*Note:* Use similar form for each job step.

changing the ways employees do their jobs. All recommendations should be discussed with employees who perform the job, and their responses should be considered carefully. If the plan is to utilize new or modified job procedures, be sure the employees understand what they are required to do and the reasons for the changes.

At this point, should we review the completed JHA? Why is a review necessary? Both questions require *yes* answers. Periodically, review of a JHA ensures that it remains current and continues to help reduce workplace accidents and injuries. Even if the job has not changed, it is possible that a review will reveal hazards not identified in the initial analysis.

Review of a JHA is critical if an illness or injury occurs on a specific job. Based on the circumstances, a review may indicate the need for a job procedure change to prevent similar incidents in the future. If an employee's failure to follow proper job procedures leads to a close call, discuss the situation with all employees who perform the job and remind them of proper procedures. Whenever a JHA is revised, it is important to train all employees affected by the changes in the methods, procedures, or protective measures.

In recent years, hazard analysis has important to many organizations. An analysis should be appropriate, applicable, and accurate, especially if the processes involved are complex. When complexity is an issue, help is available from professional consultants, the organization's insurance carrier, and the local fire department. OSHA offers assistance and consultation services through its regional and area offices (contact numbers may be found at www.osha.gov).

Organizations must understand that despite the availability of outside help, its employees *must* play a role in identifying and correcting hazards because employees are in the workplace every day and are most likely encounter the hazards. New circumstances and a recombination of existing circumstances may cause old hazards to reappear and new hazards to emerge. Management and employees must be ready, willing, and able to implement whatever hazard elimination or control measures a professional consultant recommends.

It is worth noting that OSHA can provide extensive help through a variety of safety and health programs, plans, workplace consultations, voluntary protection programs, strategic partnerships, training, education, and more. In addition to helping employers identify and correct specific hazards, OSHA's consultation service provides free on-site assistance in developing and implementing effective workplace safety and health management systems focused on preventing worker injuries and illnesses. The comprehensive assistance provided by OSHA includes a worksite hazard survey and appraisals of all aspects of existing safety and health management measures. OSHA helps employers develop and implement effective safety and health management systems. Employers also may receive training and education services and limited assistance away from their worksites.

OSHA awards grants through its Susan Harwood Training Grant Program to nonprofit organizations to provide safety and health training and education in the workplace. The grants focus on educating workers and employers in small businesses (fewer than 250 employees) and training workers and employers on new standards for high-risk activities and hazards. Grants are awarded for 1 year and may be renewed for a second year, based on satisfactory results. OSHA expects each organization awarded a grant to develop a training program that addresses a safety and health topic named by OSHA, recruits workers and managers for training, and conducts the training. Grantees are also expected to follow up with people who were trained to learn what changes were made to reduce the hazards in their workplaces as a result of the training. Each year OSHA holds a national competition. Details appear in the *Federal Register* and on the Internet (www.osha-slc.gov/Training/sharwood/sharwood.html). The OSHA Office of Training and Education is at 1555 Times Drive, Des Plaines, IL 60018, (847) 297-4810.

OSHA provides a variety of materials and tools on its website (www.osha .gov). These include eTools, expert advisors, electronic compliance assistance tools (e-CATs), technical links, regulations, directives, publications, videos, and other tools for employers and employees. OSHA's software programs and compliance assistance tools walk users through challenging safety and health issues and common problems to find the best solutions. OSHA's comprehensive publications program includes more than 100 titles explaining OSHA requirements and programs. OSHA's CD-ROM covers standards, interpretations, directives, and more and can be purchased from the U.S. Government Printing Office. To order, write to the Superintendent of Documents, U.S. Government Printing Office, Washington, DC 20402, or phone (202) 512–1800. Specify OSHA Regulations, Documents and Technical Information on CD-ROM (ORDT), GPO Order S/N729-013-00000-5.

At this point, we must emphasize that a successful safety and health program depends on management assistance. Effective management of worker safety and health protection is a decisive factor in reducing the extent and severity of work-related injuries and illnesses and their related costs. In fact, an effective safety and health program forms the basis of good worker protection and can save time and money—about $4 for every dollar spent— and increase productivity. To assist employers and employees in developing effective safety and health systems, OSHA published recommended Safety and Health Program Management Guidelines (*Federal Register* 54, 3908–3916, January 26, 1989). These voluntary guidelines can be applied to all worksites covered by OSHA and identify four general elements that are critical to a successful safety and health management program:

1. Management leadership and employee involvement
2. Worksite analysis
3. Hazard prevention and control
4. Safety and health training

The guidelines recommend specific actions based on the four elements to achieve an effective safety and health program. The *Federal Register* notice is available online. The plans were developed under the auspices of the Occupational Safety and Health Act of 1970 that encourages states to develop and operate their own job safety and health plans and permits state enforcement of OSHA standards in states that have approved plans. After OSHA approves a state plan, it funds 50% of the program's operating costs. State plans must provide standards and enforcement programs and voluntary compliance activities that are at least as effective as those of OSHA. At present, 26 state plans are in effect: 23 cover both private and public (state and local government) employment; Connecticut, New Jersey, and New York plans cover only the public sector. For more information on state plans, visit OSHA's website (www.osha.gov).

## Reference

http://www.osha.gov/Publications/osha3071.pdf

## Selected Bibliography

Bancroft, K. (2002). Job hazard analysis for unsafe acts. *Occupational Health and Safety,* 71, 206–215.

Clemens, P. and T. Pfitzer. (2006). Risk assessment and control. *Professional Safety,* 51, 41–44.

Geronsin, R. (2001). Job hazard assessment: a comprehensive approach. *Professional Safety,* 46, 23–30.

Morris, J. and J. Wachs. (2003). Implementing a job hazard analysis program. *AAOHN Journal,* 51, 187–193.

Occupational Safety and Health Administration. (2002). *Job Hazard Analysis.* Publication 3071. Washington, DC: Author.

Rozenfeld, O., R. Sacks, Y. Rosenfeld et al. (2010). Construction job safety analysis. *Safety Science,* 48, 491–498.

Swartz, G. (2002). Job hazard analysis. *Professional Safety,* 47, 27–33.

U.S. Army Corps of Engineers (2008). *Safety and Health Requirements Manual* (EM 385-1-1). Washington, DC: Government Printing Office.

U.S. Department of the Army (2010). *Army Safety Program* (385-10). Washington, DC.

U.S. Department of the Army (2006). *Composite Risk Management* (FM 100-14). Washington, DC.

# 10

## Hazard Communication Based on Standard CFR 910.1200

Many guidelines and standards are available to manufacturers to ensure that hazards are handled appropriately. This chapter will focus on the automotive industry and companies that control the hazards associated with chemicals. We will base our discussion on the CFR 1910.1200 standard focusing on chemical manufacturers and their employees.

The federal hazard communication standard (CFR 1910.1200) is based on a simple concept: employees have a need and a right to know (1) the identities and associated hazards of the chemical to which they are exposed at work and (2) the protective measures available to them at work. To ensure that this information is distributed to employees, a good approach is to utilize the standard procedures of both chemical manufacturers and employers and focus on three issues:

1. Improved control of chemicals
2. Promotion of safe and healthy work practices
3. Fostering ability to recognize, evaluate, and control chemical hazard potential

We first review the responsibilities chemical manufacturers and employers. Chemical manufacturers must evaluate the hazards of all chemicals they produce to determine effects on the health and safety of people exposed to them. Chemical manufacturers must provide information on all their products that:

- Have been shown to cause adverse health effects.
- Have the potential to cause adverse health effects.
- Can cause physical hazards, such as fire or explosion.
- Require additional health hazard information.

Chemical manufacturers also must evaluate whether as chemical is flammable, combustible, an oxidizer, explosive, unstable, or water reactive. Employer responsibilities are key components of the hazard communication standard. Every employer whose employees may be exposed to chemical materials must maintain a written hazard communication program that includes (1) provisions for informing employees about hazards and (2) training to help them understand the hazard information. To meet these general requirements, the written program must include:

- Labeling system ensuring that each chemical container is labeled, tagged, or marked to show its identity and appropriate hazard warnings.
- Material safety data sheets (MSDSs) to ensure that technical information (ingredients, hazards, and guidelines for use) for all chemicals used or stored in the workplace are is available to employees.
- Employee information and training providing effective information and training on chemicals utilized in their work areas.
- Safe use instructions (SUIs) containing specific instructions on the safe use and handling of chemicals in work areas.
- Chemical materials lists detailing all chemicals; lists should be updated as required.
- Non-routine tasks may involve chemical hazards; employees who perform such tasks must be notified of the hazards.
- Pipes and piping systems also involve chemical-associated hazards; employees who work with unlabeled pipes must be informed of related hazards.
- Access to information about hazardous materials must be provided to employees who work with or are exposed to them.
- On-site contractors must be informed about chemical materials to which their employees may be exposed. Also, such contractors must supply the company with MSDSs for all chemicals they deliver to the facility.

## Hazard Communication Program and Hazardous Materials Control Committee

Most organizations dealing with hazards maintain some forms of communication programs that generally contain guidelines to ensure all areas meet the federal hazardous communication standard. However, facilities vary considerably and implementations of programs also vary from plant to plant. All programs, however, contain the same basic elements. The first is a hazardous materials control committee (HMCC) that serves as the foundation of a hazard communication program. Every facility has an HMCC.

### Members

HMCC members are workers who may come from engineering, medical, health and safety, production, or research areas. If an operation is unionized, a health and safety technician and a union industrial hygiene representative will also be members. The HMCC also obtains advice from experts in safety, industrial hygiene, firefighting, toxicology, and medicine.

### Responsibilities

The major task of the HMCC is to ensure the success of the hazardous communication program. The committee is responsible for (1) developing a written hazard communication program specific to the facility and (2) approving all chemical materials and processes. The committee uses MSDSs to decide whether new chemicals are safe. If the HMCC does not approve a chemical, it may not be used.

### Employee Training

An important part of a hazard communication program is employee training. Employees learn:

- To recognize potential physical and health hazards of chemicals in their work areas
- To use chemical materials safely
- About potential hazards of chemical materials in pipes or piping systems
- About the Occupational Safety and Health Administration (OSHA) standard
- Where to find labels, MSDSs, SUIs, and chemical materials lists and how to use them

Additional training may be required if (1) new physical and/or health hazards for which employees have not been previously trained are introduced; (2) existing chemicals are used in a new way; (3) employees are assigned new jobs that involve chemical for which they have not been trained; (4) an employee performs a non-routine (rarely performed) task; (5) new hazard information about a chemical material becomes available. A typical safe use training program covering chemicals may include the following:

- Program overview
- Understanding hazards
- Detecting and evaluating hazards
- Controlling hazards
- Safe use category system
- Halogenated solvents
- Solvents with flashpoints below 100°F

- Solvents with flash points exceeding 100°F
- Metalworking fluids and lubricants
- Adhesives such as solvent-based polyurethanes, epoxies, and sealers
- Adhesives and sealers
- Toxic compressed gases
- Metals, metal salts, metal powders, and solders
- Corrosive concentrated acids with pH values less than 4
- Corrosive acid or base powders, flakes, and salts
- Corrosive concentrated bases (alkalis) with pH values exceeding 10
- Flammable compressed gases
- Inert compressed gases
- Oxidizing compressed gases
- Corrosive compressed gases
- General use
- Health risks
- Unique hazards
- Respirable and reactive fibers and particulates
- Biocides and pesticides
- Oxidizing materials

The first five topics generally constitute an overview of the hazardous communication program. They cover various ways chemicals can be hazardous and how hazards can be controlled and explain the safe use category system. All employees must receive training on the first five topics as a minimum. The next 20 items provide information on each safe use category, including explanations of the categories, possible hazards, and safe use. Employees receive training on the categories specific to the chemicals to which they may be exposed. For the training to be successful, all employees must participate.

### Employee Access

The third element of a hazard communication program is employee access. All employees who work with or have potential exposure to chemicals have the right to access to MSDSs, safe use instructions, chemical materials lists, and the written hazard communication program.

Any employee working with a potential hazard has the right to review SUIs at all times while in the work area. He or she does not need a written request to review and discuss the information in MSDSs, chemical materials lists, or the written hazard communication program. He or she may make a written request for copies of these documents. The area supervisor or union representative should know where to find this information.

## Information Sources

The final element of a hazard communication program is the availability of various forms of information to employees who work with or near chemicals: labels, SUIs, chemical; materials lists, and MSDSs.

### Labels

All chemical containers must be labeled. Bags, barrels, bottles, boxes, cans, cylinders, drums, reaction vessels, and storage tanks are all containers. Storage tanks include tank trucks and railcars. Labels are used to identify a chemical and provide appropriate hazard warnings. Every label must clearly state the material identity (name or number by which the chemical is known in the facility). The identity noted on the label should match the product name on the MSDS and list of chemical materials.

Every label must include a warning describing the health and physical effects of overexposure to the chemical and cover effects on specific organs. The warning may consist of or combine words, pictures, and symbols to convey the required information.

If an operation transfers small quantities of chemicals from large drums to smaller containers, employees must use dedicated transfer containers with permanent labels. However, if a worker uses the chemical from the large container immediately (during the current shift), it is not necessary to have a dedicated transfer container or permanent label. In certain types of operations, dedicated containers are always required and the time of use does not matter.

Under certain conditions, alternative labeling on chemical containers is appropriate. Signs, placards, process sheets, and/or batch tickets may be used in place of labels on individual stationary process containers. Alternative labels must contain the required information and attached so that they cannot be removed easily. Table 10.1 is a sample label.

### Safe Use Instructions

Safe use instructions (SUIs) provide more information than labels. They list specific instructions for the safe use and handling of a chemical. They are simpler, easier-to-read versions of MSDSs. A hazardous materials control system (HMCS) is usually an electronic database facility dependent that generates SUIs and attachments for chemical materials in the safe use category system. Attachments may list specific chemical names and additional health hazard information.

The information on SUIs is based on how a chemical is used in a specific job. If the same chemical is used in more than one operation, individual SUIs may be required to cover each operation. SUIs for all chemicals used in a department should be maintained in reference manual available to employees and supervisors. A typical SUI will indicate:

**TABLE 10.1**

Chemical Container Label

| Item | Description | Identification |
|------|-------------|----------------|
| 1 | Product name or material identity | |
| 2 | If solvent, flash point > 100°F | |
| | Safe use category[a] | |
| 3 | The < symbol means less than; > means greater than | |
| 4 | WARNING! Overexposure may result in central nervous system (CNS) effects, including headache, dizziness, nausea, unconsciousness, death. Skin irritant. Possible liver and kidney effects | |
| 5 | Check appropriate lines below: | |
| | ☐ Do not use in confined space without appropriate personal protective equipment (PPE) | |
| | ☐ Flammable | |
| 6 | Health hazards: | |
| | ☐ Harmful if inhaled or swallowed | |
| | ☐ Harmful if absorbed through skin | |
| | ☐ Cancer-suspect agent | |
| 7 | Specific chemicals with additional health hazards. See safe use instructions | |

Legend:
1. Names or numbers of chemical materials as used in facility
2. Group of chemicals (safe use category) to which material belongs
3. The < symbol means *less than*; > means *greater than*
4. Effects of overexposure
5. Special precautions
6. How material enters the body and whether it contains a cancer-suspect agent
7. Ingredients that need special precautions for use
[a] Employees have the right and are encouraged to review MSDSs and SUIs to obtain additional chemical and health hazard information.

1. Name or number of the chemical material as it is used in the operation
2. Group of chemicals (safe use category) to which the material belongs
3. How the material enters the body
4. What happens if you are overexposed
5. What must be done in an emergency
6. How to handle the material safely
7. Equipment required to protect workers from exposure
8. What to do if a fire occurs
9. How the chemical reacts with other chemicals
10. What to do if spills or leaks occur
11. How to store the material safely
12. How to discard the material safely
13. Additional instructions for specific uses

Certain SUIs will include attachments that list specific chemical ingredients that need special precautions or are used widely. Table 10.2 is a typical SUI showing the points mentioned.

### Chemical Materials Lists

The hazard communication standard requires that employers maintain lists of hazardous chemicals present in their workplaces. Chemicals listed must be identified by product name or material identity. This identification must match the MSDS identification.

The list should be kept in a central location (e.g., safety, security, or medical department) and employees and supervisors should be advised of the location. Individual departments may also maintain lists of chemicals they use. If your facility maintains departmental lists, copies should be included in the relevant employee and supervisor reference manual.

### Material Safety Data Sheets

MSDSs provide technical information about chemicals. They list ingredients, hazards, and general guidelines for use. Typically, every manufacturer that sells a chemical to an organization must provide an MSDS *before* shipment. Usually, a technical review staff of a company's chemical risk management group examines MSDSs to verify that they are complete or ensure that missing information is added.

MSDSs may be stored electronically in a hazardous materials control system (HMCS) and should be updated whenever chemicals are introduced, modified, or discontinued. Each facility determines where master sets (hard copies) of the MSDSs are stored and the location should be known by all employees and supervisors. Table 10.3 is a generic version of a MSDS.

1. Chemical product and manufacturer identification
2. Composition and list of ingredients
3. Hazards identifications
4. First aid measures
5. Fire fighting measures
6. Accidental release measures
7. Handling and storage procedures
8. Exposure controls and protective measures
9. Physical and chemical properties
10. Stability and reactivity
11. Toxicological data
12. Ecological data
13. Disposal instructions

**TABLE 10.2**

Typical SUI Form

| Chemical Risk Management | Safe Use Instruction<br>Print Date:<br>Page ___ of ___ | Creation Date:<br>Safe Use Category (SUC): |
|---|---|---|

**Location**

Department:

Work area:

Process:

Occupation:

**(1) Material Identification**

Product name:

Manufacturer:

Supplier:

Hazardous ingredients:

**(2) Health Hazards**

Inhalation: Central nervous system effects (headache, dizziness, nausea, loss of consciousness, brain damage, death).

Eye and skin irritation: Dry or cracked skin, rash, redness, burning, itching. Skin absorption may be harmful.

Ingestion: Gastrointestinal disturbance. Possible liver and kidney effects.

Additional comments of chemical producer:

Exposure may be by inhalation and/or skin or eye contact, depending on conditions of use. To minimize exposure, follow recommendations for proper use, ventilation, and personal protective equipment.

Effects of overexposure include irritation of upper respiratory system. May cause nervous system depression. Extreme overexposure may cause loss of consciousness and death.

Signs and symptoms of overexposure to vapors or spray mists include headache, dizziness, nausea, and loss of coordination.

Chronic health hazards: prolonged overexposure to solvent ingredients may cause adverse effects to the liver, urinary, and cardiovascular systems.

Notes: Intentional misuse by deliberately concentrating and inhaling contents may be harmful or fatal. Reports associate repeated and prolonged overexposure to solvents with permanent brain and nervous system damage.

**(3) Emergency First Aid**

First aid phone:

Emergency location:

Effects of exposure and emergency first aid measures:

After skin contact: Wash affected area thoroughly with soap and water. Remove contaminated clothing and launder before reuse.

After eye contact: Flush eyes with large amounts of water for 15 minutes. Get medical attention.

After inhalation: Remove from exposure. Restore breathing. Keep warm and quiet.

After ingestion: Do not induce vomiting. Get medical attention immediately.

**TABLE 10.2 (Continued)**

Typical SUI Form

**Protection Measures and Procedures**

**(4) Material Use Instructions**

Do not use in confined areas without first consulting confined space procedure. Breathing vapors may he hazardous. Ventilation may be required, where feasible, to maintain concentrations below applicable limits (OSHA guidelines).

Avoid contact with skin and eyes. Keep container tightly closed when not in use or empty. Keep container away from heat, sparks, and open flames. Container must be grounded when in use and bonded when contents are transferred. Use approved safety containers.

Wash hands after use and before eating, drinking, smoking, or applying cosmetics. No smoking allowed in area of use. Clothing soaked through to skin should he removed and laundered before reuse.

Engineering controls: Ventilation, local exhaust preferable. General exhaust acceptable if exposure to material is maintained below applicable exposure limits. See OSHA Standards 1910.94, 1910.107, and 1910.108.

Plant instructions: [use this field for plant-specific data]

**(5) Personal Protective Equipment**

Eye protection: Wear safety glasses with unperforated side shields.

Respiratory protection: If personal exposure cannot be controlled below applicable limits by ventilation, wear properly fitted organic vapor or particulate respirator approved by NIOSH/MSHA for protection. Wear nitrile rubber or silver shield gloves recommended by supplier for protection against materials.

Plant Instructions: Use maximum recommended protection.

**(6) Reactivity Information**

**(7) Storage Instructions**

Contents are flammable. Keep away from heat, sparks, and open flame during use and until all vapors are gone. Keep area ventilated. Do not smoke. Extinguish all flames, pilot lights, and heaters. Turn off stoves, electric tools, appliances, and other sources of ignition. Consult NFPA code.

Use approved bonding and grounding procedures. Keep container closed when not in use. Transfer contents only to approved containers with complete and appropriate labeling. Do not take internally. Keep out of the reach of children.

Use only with adequate ventilation. Avoid contact with skin and eyes. Avoid breathing vapor and spray mist. Wash hands after using.

**Instructions in Case of Danger**

Fire emergency phone:

**(8) Fire Fighting Instructions**

Extinguishing media: Carbon dioxide, dry chemical, and foam. Class B extinguishing media not suitable.

Special procedures: Full protective equipment including self-contained breathing apparatus should be used. Water spray may be ineffective. If water is used, fog nozzles are preferable. Water may be used to cool closed containers to prevent pressure build-up and possible auto-ignition or explosion from exposure to extreme heat. Closed containers may explode when exposed to extreme heat.

*Continued*

**TABLE 10.2 (Continued)**

Typical SUI Form

Application to hot surfaces requires special precautions. During emergency conditions, overexposure to decomposition products may cause a health hazard. Symptoms may not be immediately apparent. Obtain medical attention.

**(9) Spill and Leak Instructions**

Remove all sources of ignition. Ventilate area.

Small spills (less than 1 gallon): contain spill and keep from entering sewer. Protect drain by diking. Remove spill with inert absorbent. Contact supervisor.

Large spills (more than 1 gallon): Notify security or follow local emergency response procedure. Do not smoke.

**(10) Plant Instructions**

Contact plant environmental engineer at Ext. \_\_\_\_ to report spills.

**(11) Other Dangers**

If material is packaged as an aerosol, keep away from heat, sparks, open flames, and direct sunlight. Contents are under pressure. Do not puncture can. In high heat conditions, aerosol cans may vent, rupture, rocket, or explode.

**(12) Disposal Instructions**

Waste from this product may be hazardous as defined by the Resource Conservation and Recovery act (RCRA), 40 CFR 261. Waste must he tested for ignitability to determine applicable EPA hazardous waste numbers. Incinerate in approved facility. Do not incinerate closed container. Dispose of in accordance with federal, state or provincial, and local pollution regulations.

Remove waste with inert absorbent. Contact plant resource management for disposal instructions.

**(13) SUI Attachment**

| | | |
|---|---|---|
| SUC No. | | |
| Chemical name: | | |
| Product name: | | |
| HMCS ID: | | |
| Component chemicals: | | |

**TABLE 10.3**

Generic Material Safety Data Sheet

| HMCS Identification | SUC: |
|---|---|
| Product Name: | Supplier: |
| Manufacturer: | Original date: |
| | Revision date: |
| | Effective date: |
| | Print date: |
| | Page ___ of ___ |

Product and company identification **(1)**:

Information about substance and/or preparation:

Information about manufacturer:

Information about supplier:

Composition and pertinent information about ingredients **(2)**:

Hazard identification (practical observations) **(3)**:
Exposure routes:

Specific hazards and their effects:

Emergency overview:

Aggravation by preexisting medical conditions:

Additional comments:

First aid measures **(4)**
First aid for:
   Inhalation:
   Skin contact:
   Eye contact:
   Ingestion:
   General:

Firefighting measures **(5)**
Product flammability:

Suitable extinguishing media:
Fire and explosion hazards:

Special firefighting methods:

Additional comments:

*Continued*

**TABLE 10.3 (Continued)**

Generic Material Safety Data Sheet

Accidental release measures **(6)**
Precautions in case of accidental release

    Personal precautions:

Cleanup methods:

    Neutralization:

Handling and storage **(7)**
Handling

    Safe-handling advice:

    Prevention of fire and explosion:

Storage

    Technical measures and/or storage conditions:

    Storage class and other classes:

Exposure control and personal protection **(8)**

Engineering measures:

Exposure limits:
    Limit values for all chemicals

Personal protective equipment (PPE):
    Skin protectors
    Eye protectors
    Hearing protectors
    Respiratory protectors
    Appropriate protective clothing (gloves, uniforms, etc.)
Hygiene measures:

Physical and chemical properties **(9)**
Appearance:
    Physical state (liquid, solid, etc.)
    Form
    Color
    Odor

**TABLE 10.3 (Continued)**

Generic Material Safety Data Sheet

Relevant safety data:
  pH value
  Changes of state
  Flash point
  Flammability
  Explosive limits
  Vapor pressure
  Vapor density
  Evaporation rate
  Density
  Specific gravity
  Solubility
  Volatile organic compounds (VOCs):
Additional chemical and physical data:

Stability and reactivity **(10)**
  Stability under normal conditions:

  Conditions to avoid:

  Mechanical impact stability:

  Conditions causing hazardous polymerization:

Hazardous decomposition products:

Additional comments:

Toxicological information **(11)**

Practical experiences:
  Health effects:

Classification of ingredients
  Carcinogenicity:

  General remarks:

  Additional comments:

Ecological information **(12)**

Environmental impact and comments:

Ecotoxicity and comments:

*Continued*

**TABLE 10.3 (Continued)**

Generic Material Safety Data Sheet

Disposal considerations **(13)**
Additional comments:

Transport information **(14)**
DOT information

Regulatory information **(15)**
  National regulations:
  SARA 311/312:
  Immediate health:
  Delayed health:
  Fire:
  Sudden pressure release:
  Reactive:
  Resource Conservation and Recovery Act (RCRA):

Hazard waste:

Additional comments:

Other information **(16)**

Additional comments:

Changes:

14. Transport information
15. Regulatory requirements
16. Other information about the product that may be useful to the MSDS user

# References

29 CFR 1910.1200. Hazard Communication.
https://www.osha.gov/pls/oshaweb/owadisp.show_document?p_table=STANDARDS&p_id=10100
U.S. Department of Labor. *OSHA*. Occupational Safety & Health Administration. Washington. www.OSHA.gov

# Appendix A: Checklists

This appendix provides a variety of useful checklists. They are by no means exhaustive. Rather they serve as guidelines for developing operation-specific checklists. Readers should always remember that there is no such thing as a complete or perfect checklist. Some items that appear on a checklist should not be covered in meetings because they may not be relevant to the discussion. Conversely, items not identified on a checklist may require discussion.

## Safety Plan Checklist

Generating an effective safety checklist involves two main principles: (1) good safety management and (2) a basic safety philosophy. The basic requirements of good safety management are:

- Management commitment
- Documented safety philosophy
- Safety goals and objectives
- In-house safety committee
- Line responsibility for safety
- Supportive safety staff
- Rules and procedures
- Audits
- Safety communications
- Safety training
- Accident investigation
- Motivation

The basic philosophy is based on the principles that:

- Every incident can be avoided
- No job is worth getting hurt for
- Every job will be done safely
- Incidents can be managed
- Safety is everyone's responsibility

- Safety and best manufacturing practices must be followed.
- Safety standards, procedures, and practices must be developed.
- Everyone must understand and meet all training requirements.
- Working safely is a condition of employment.

After the requirements and philosophy are understood and management is ready to implement them into a plan, a checklist may be generated. Table A.1 is a checklist showing desired elements for safety plans based on hydrogen and fuel cells. The left column cites page numbers in the document as a cross-reference to help project teams verify that their safety plan is complete and can be a valuable tool over the life of a project as a cross-reference. Duplicate page numbers indicate multiple concerns.

---

### Facility Location Checklist

Table A.2 is a typical checklist for a facility review. It identifies various types of hazard concerns.* It differs from Table A.1 in that it lists hazards by location or plant activity.

---

### Australian Health Administration (AHA) Guidelines and Checklist

Undertaking an AHA evaluation requires consideration of specific issues. The team should develop several guidelines in the form of basic questions to evaluate each documented hazard. Some of the common guideline questions are:

- Have all possible steps of the operation been described?
- Are all potential hazards identified?
- Are the controls sufficient for the identified hazards?
- Is the evaluation clear and concise?
- Is the information accessible to all employees?

---

* It is based on http://www.epa.state.oh.us/portals/27/112r/facilitysiting.pdf; http://www.werc. org/assets/1/workflow_staging/Publications/434.PDF; and http://www.ccohs.ca/oshanswers/ hsprograms/list_mft.html

**TABLE A.1**

Safety Plan Checklist

| Page | Element | The Safety Should Describe |
|---|---|---|
| 1 | Scope of work | Nature of the work being performed |
| 3 | Organizational policies and procedures | Application of organizational safety-related policies and procedures for work being performed |
| 3 | Hydrogen and fuel cell experience | How previous organization experience with hydrogen, fuel cells, and related work applies to project |
| 4 | Identification of safety vulnerabilities (ISV) | ISV methodology applied to project FMEA, what-if, HAZOP, checklist, FTA, ETA, PRA, or other method |
| | | Designation of leader and steward of ISV methodology |
| | | Significant accident scenarios identified |
| | | Significant vulnerabilities identified |
| | | Safety-critical equipment determined |
| | | Storage and handling of hazardous materials and ignition sources |
| | | Explosion hazards and material interactions |
| | | Leakage and accumulation detection |
| | | Hydrogen handling systems (supplies, storage, distribution, volumes, pressures, estimated use rates) |
| 4 | Risk reduction plan | Prevention and mitigation measures for significant vulnerabilities |
| 4 | Operating procedures | Operational procedures applicable to location and performance of work including sample handling and transport |
| | | Operating steps that must be written to show critical variables, and acceptable ranges and responses to deviations |
| 5 | Equipment and mechanical integrity | Initial testing and commissioning |
| | | Preventative maintenance plan |
| | | Calibration of sensors |
| | | Test and inspection frequency and documentation |
| 6 | Management of change procedures | System and/or procedures for reviewing proposed changes of materials, technology, equipment, procedures, staffing, and facility operation to determine effects on safety vulnerabilities |
| 6 | Project safety documentation | Communication of required safety information that must be available to all project participants. Safety information includes ISV documentation, procedures, handbooks, standards, other references, and safety review reports |
| 7 | Employee training | Requirement for initial and refresher general safety training |
| | | Initial and refresher hydrogen-specific and hazardous material training |
| | | System allowing organization stewards to participate in training participation and confirm their understanding |

*Continued*

**TABLE A.1 (Continued)**

Safety Plan Checklist

| Page | Element | The Safety Should Describe |
|---|---|---|
| 7 | Safety reviews | Applicable safety reviews beyond ISV described above |
| 7 | Safety events and lessons learned | Procedure for reporting to management and U.S. Department of Energy (DOE) on: System and/or procedure for investigating events Method for implementing corrective measures Document and dissemination of lessons learned from incidents and near-misses |
| 9 | Emergency response | Plans and procedures for responding to emergencies Communication and interaction with local emergency response officials |
| 9 | Self-audit | System for verifying that safety-related procedures and practices are followed throughout life of project |
| 9 | Safety plan approval | Documented safety plan review and approval process |
| 9 | Other comments or concerns | Information on topics not covered above Issues that may require DOE assistance |

*Source:* http://www1.eere.energy.gov/hydrogenandfuelcells/pdfs/safety_guidance.pdf

Based on the answers to the basic questions above, a checklist may be used as a guide for an AHA evaluation and reveal safety deficiencies. Critical questions for a checklist are:

- Is work activity stated?
- Is the project name stated?
- Is preparer named?
- Is date of completion noted?
- Is the AHA more than 2 years old?
- Does more than one AHA cover the same task?
- If so, are they comparable?
- Does the AHA need revalidation?
- Are checklists and data centrally located for access?
- Have potential hazards been missed?
- Are tools and equipment listed and described?
- Is appropriate documentation attached?
- Does the project involve Army Corps–specific activity?
- Do the job steps need to be clarified?

**TABLE A.2**

Facility Location Checklist

| Area of Concern or Question | Response | Recommendations |
|---|---|---|
| *Spaces between Process Components* | | |
| 1. Have adequate provisions been made for relieving explosions in process equipment? | | |
| 2. Are operating units and machines within units spaced to minimize potential damage from fires or explosions in adjacent areas? | | |
| 3. Are there safe exit routes from each unit? | | |
| 4. Has equipment been adequately spaced and located to safely permit anticipated maintenance (pulling heat exchanger bundles, dumping catalyst, crane lifting) and hot work? | | |
| 5. Are vessels containing highly hazardous chemicals located sufficiently far apart? If not, what hazards are introduced? | | |
| 6. Is there adequate access for emergency vehicles such as fire trucks? | | |
| 7. Can adjacent equipment or facilities withstand the overpressure generated by explosions? | | |
| 8. Can adjacent equipment and facilities such as support structures withstand flame impingement? | | |
| *Locations of Large Inventories* | | |
| 1. Are large inventories of highly hazardous chemicals located away from process areas? | | |
| 2. Is temporary storage provided for raw materials and finished products at appropriate locations? | | |
| 3. Are inventories of highly hazardous chemicals held to a minimum? | | |
| 4. Are reflux tanks, surge drums, and rundown tanks located to prevent concentrations of large volumes of highly hazardous chemicals in any one area? | | |
| 5. Has special consideration been given to storing and transporting explosives? | | |
| 6. Have the following items been considered in locating material handling areas: <br> • Fire hazards? <br> • Proximity to important buildings? <br> • Safety devices such as sprinklers? <br> • Slope; is the area level? | | |

*Continued*

**TABLE A.2 (Continued)**

Facility Location Checklist

| Area of Concern or Question | Response | Recommendations |
|---|---|---|
| *Location of Motor Control Center (MCC)* | | |
| 1. Is the MCC easily accessible to operators? | | |
| 2. Are circuit breakers easy to identify? | | |
| 3. Can operators safely open circuit breakers? Have they been trained to do so? | | |
| 4. Is the MCC designed NOT to be an ignition source? Are doors always closed? Is no-smoking policy strictly enforced? | | |
| 5. Is the MCC designed to be a safe haven? | | |
| *Location and Construction of Control Rooms* | | |
| 1. Is the control room built to satisfy current over-pressure and safe-haven standards? | | |
| 2. Does the construction of the control room satisfy all relevant criteria such as factor mutual recommendations? | | |
| 3. Are control room workers and their escape routes protected from: | | |
| • Toxic, corrosive, or flammable sprays, fumes, mists, or vapors? | | |
| • Thermal radiation from fires, including flares? | | |
| • Over-pressure and projectiles from explosions? | | |
| • Contamination from spills or runoff? | | |
| • Noise? | | |
| • Contamination of utilities such as breathing air? | | |
| • Transport of hazardous materials from other sites? | | |
| • Possible long-term worker exposure to low concentrations of process materials? | | |
| • Odors? | | |
| • Impacts (e,g., from a forklift)? | | |
| • Flooding (e.g., ruptured storage tank)? | | |
| 4. Are vessels containing highly hazardous chemicals located sufficiently far from the control room? | | |
| 5. Were the following characteristics considered when the control room location was determined: | | |
| • Construction materials and methods? | | |
| • Types and quantities of materials? | | |

**TABLE A.2 (Continued)**

Facility Location Checklist

| Area of Concern or Question | Response | Recommendations |
|---|---|---|
| • Directions and velocities of prevailing winds? | | |
| • Types of reactions and processes? | | |
| • Operating pressures and temperatures? | | |
| • Ignition sources? | | |
| • Fire protection facilities? | | |
| • Drainage facilities? | | |
| 6. Are windows of rigid construction with sturdy panes (e.g., of woven-wire reinforced glass)? | | |
| 7. Is at least one exit located in a direction away from the process area? Do exit doors open outward? Are emergency exits provided for multistory control buildings? | | |
| 8. Do the ends of horizontal vessels face away from the control rooms? | | |
| 9. Are critical pieces of equipment in the control room well protected? Does the control room have adequate barricading? | | |
| 10. Are open pits, trenches, or other pockets where inert, toxic, or flammable vapors may collect located away from control buildings or equipment that utilizes flammable fluids? | | |
| 11. Are pipes, wires, and conduits sealed at the points where they enter the building? Is the building sealed at the point of entry? Have other potential leakage points into the building been adequately sealed? | | |
| 12. Is the control room located a sufficient distance from sources of excessive vibration? | | |
| 13. Is a positive pressure maintained in control rooms in hazardous areas? | | |
| 14. Could any structure fall on the control room in the event of an accident? | | |
| 15. Is the roof of the control room free from heavy equipment and machinery? | | |

*Locations of Machine Shops, Welding Shops, Electrical Substations, Roads, Rail Spurs, and Other Likely Ignition Sources*

| | | |
|---|---|---|
| 1. Are likely ignition sources (maintenance shops, roads, rail spurs) located away from release points for volatile liquids and vapors? | | |
| 2. Are process sewers located away from likely sources of ignition? | | |

*Continued*

**TABLE A.2 (Continued)**

Facility Location Checklist

| Area of Concern or Question | Response | Recommendations |
|---|---|---|
| 3. Are all vessels containing highly hazardous chemicals containing material above their flash points located away from likely sources of ignition? | | |
| 4. Are flare and fired heater systems located to minimize hazards to workers and equipment based on normal wind direction and velocity and heat potential? | | |

*Locations of Engineering, Laboratories, Administration, and Other Structures*

| | | |
|---|---|---|
| 1. Are administration buildings located away from inventories of highly hazardous chemicals? | | |
| 2. Are administration buildings located away from release points of highly hazardous chemicals? | | |
| 3. Are workers in administration buildings protected from: | | |
|   • Toxic, corrosive, or flammable sprays, fumes, mists, or vapors? | | |
|   • Thermal radiation from fires, including flares? | | |
|   • Over-pressure and projectiles from explosions? | | |
|   • Contamination of utilities such as water? | | |
|   • Contamination from spills or runoff? | | |
|   • Noise? | | |
|   • Transporting hazard materials from other sites? | | |
|   • Flooding (e.g., ruptured storage tank)? | | |

*Unit Layout*

| | | |
|---|---|---|
| 1. Are large inventories or release points of highly hazardous chemicals located away from vehicular traffic within the plant? | | |
| 2. Could specific siting hazards arise from external forces such as high winds, earth movement, outside utility failures, flooding, natural fires, and fog? | | |
| 3. Do emergency vehicles such as fire trucks have clear access? Are access roads free from blockage from trains and highway congestion? | | |
| 4. Are access roads engineered to avoid sharp curves? Are traffic signs provided? | | |

**TABLE A.2 (Continued)**

Facility Location Checklist

| Area of Concern or Question | Response | Recommendations |
|---|---|---|
| 5. Is vehicular traffic restricted from areas where pedestrians could be injured or equipment damaged? | | |
| 6. Are cooling towers located so that the fog they generate will not be a hazard? | | |
| 7. Do the ends of horizontal vessels face away from personnel areas? | | |
| 8. Is hydrocarbon-handling equipment located outdoors? | | |
| 9. Are pipe bridges located not to extend over equipment, control rooms, and administration buildings? | | |
| 10. Is piping design adequate to withstand potential liquid loads? | | |

*Location of Unit Relative to On-Site and Off-Site Surroundings*

| | | |
|---|---|---|
| 1. Is a system in place to notify neighboring units, facilities, and residents if a release occurs? | | |
| 2. Are detection systems and/or alarms in place to assist in warning neighboring units, facilities, and residents if a release occurs? | | |
| 3. Do neighbors (including units, facilities, and residents) know how to respond when notified of a release? Do they know how to shelter in place and when to evacuate? | | |
| 4. Are large inventories or release points for highly hazardous chemicals located away from public roads? | | |
| 5. Is the unit located or can it be relocated to minimize off-site or intra-site transportation of hazardous materials? | | |
| 6. Are workers in adjacent units and plants and the public and environmental receptors protected from the release of highly hazardous chemicals such as those listed below? | | |
|   • Toxic, corrosive, or flammable sprays, fumes, mists, or vapors? | | |
|   • Thermal radiation from fires, including flares? | | |
|   • Over-pressure from explosions? | | |
|   • Contamination from spills or runoff? | | |
|   • Odors? | | |
|   • Contamination of utilities such as sewers? | | |

*Continued*

**TABLE A.2 (Continued)**

Facility Location Checklist

| Area of Concern or Question | Response | Recommendations |
|---|---|---|
| • Transporting hazardous materials from other sites? | | |
| • Impacts such as airplane crashes and derailments? | | |
| • Flooding (e.g., ruptured storage tank)? | | |
| 7. Are workers protected from the effects of adjacent units or facilities that impact all the items listed above? | | |

*Locations of Firewater Mains and Backup Pumps*

| | | |
|---|---|---|
| 1. Are firewater mains easily accessible? | | |
| 2. Are firewater mains and pumps protected from over-pressure and blast debris impacts? | | |
| 3. Is an adequate water supply available for firefighting? | | |
| 4. Are firehouse doors pointed away from the process area so they will not be damaged by explosion over-pressure? | | |

*Location and Adequacy of Drains, Spill Basins, Dikes, and Sewers*

| | | |
|---|---|---|
| 1. Are spill containments sloped away from process inventories and potential fire sources? | | |
| 2. Have precautions been taken to avoid open ditches, pits, sumps, or pockets where inert, toxic, or flammable vapors could collect? | | |
| 3. Are process sewers transporting hydrocarbon materials closed systems? | | |
| 4. Are concrete bulkheads, barricades, or berms installed to protect workers and equipment from explosion and/or fire hazards? | | |
| 5. Are vehicle barriers installed to prevent impacts to critical equipment near high-traffic areas? | | |
| 6. Do drains empty to areas where materials cannot pool? | | |
| 7. Can dikes hold the capacity of the largest tank? | | |
| 8. Is there a means of access in and out of dikes, pits, etc.? | | |

*Locations of Emergency Stations (Showers, Respirators, PPE, etc.)*

| | | |
|---|---|---|
| 1. Are emergency stations easily accessible? | | |
| 2. Are first aid stations well located and adequately equipped? | | |

**TABLE A.2 (Continued)**

Facility Location Checklist

| Area of Concern or Question | Response | Recommendations |
|---|---|---|
| 3. Are safety showers heated, freeze protected, and wind protected? | | |
| 4. Is there a control room alarm? | | |

*Electrical Classification*

| | | |
|---|---|---|
| 1. Is there an electrical classification document? | | |
| 2. Does the document appear correct and complete? | | |
| 3. Has the document been revised recently? | | |
| 4. Have significant changes made since system construction been explained in the electrical classification document: | | |
|   • New materials added? | | |
|   • New sources of flammable gases or vapors? | | |
|   • New low points (sumps or trenches) at grade? | | |
|   • Areas that have been enclosed since the system was constructed? | | |
| 5. Are the designs and maintenance programs for ventilation systems adequate: ventilation systems being properly maintenance, and alarms and interlocks on these systems periodically function checked? | | |
|   • Regular maintenance to check functioning of natural ventilation systems? | | |
|   • Technical bases for design changes to ventilation system? | | |
|   • Ventilation systems verified adequate for new gas or vapor loads? | | |
| 6. Will safeguards alert operators when a ventilation system fails? | | |
| 7. Are controls adequate to ensure that electrically qualified equipment is replaced with equipment of equal or higher classification? | | |
| 8. Are physical boundaries in place between electrically classified areas? If not: | | |
|   • Are boundaries marked? | | |
|   • Do workers understand the boundaries of electrically classified areas and their importance? | | |
| 9. Are Division 1 areas necessary? | | |

*Continued*

**TABLE A.2 (Continued)**

Facility Location Checklist

| Area of Concern or Question | Response | Recommendations |
|---|---|---|
| 10. Are controls (e.g., hot work permit system) covering repair and construction activities? Do they include work by contractor personnel? | | |
| 11. Does the electrical classification adequately show the effects of various modes of operation (normal, maintenance, start-up, and infrequent operating modes such as reactor regeneration or operation with a portion of the system bypassed)? | | |

*Contingency Planning*

| | | |
|---|---|---|
| 1. What expansion or modification plans does the facility have? | | |
| 2. Can the unit be built and maintained without transporting heavy items above operating equipment and piping? | | |
| 3. Are calculations, charts, and other documents available to verify that facility location has been considered in the layout of the unit? Do these documents show that consideration has been given to: | | |
| • Normal direction and velocity of wind? | | |
| • Atmospheric dispersion of gases and vapors? | | |
| • Estimated radiant heat density created by a fire? | | |
| • Estimated over-pressure? | | |
| 4. Are appropriate security safeguards (fences and guard stations) in place? | | |
| 5. Are gates located away from public road so that the largest trucks can move completely off the roadway while waiting for gates to be opened? | | |
| 6. Where applicable, are safeguards in place to protect high structures against low-flying aircraft? | | |
| 7. Are adequate safeguards in place to protect employees against exposure to excessive noise? Do safeguards consider the cumulative effects of machines located in close proximity? | | |
| 8. Is adequate emergency lighting provided? Is adequate back-up power available for this lighting? | | |
| 9. Are procedures in place to restrict non-essential or untrained personnel from entering areas deemed hazardous? | | |

**TABLE A.2 (Continued)**

Facility Location Checklist

| Area of Concern or Question | Response | Recommendations |
|---|---|---|
| 10. Are indoor safety control systems (fire walls and sprinklers) installed in buildings housing control rooms and administrative functions? | | |
| 11. Are evacuation plans from buildings and areas adequate and accessible to workers? | | |
| 12. Are evacuation drills conducted routinely? | | |

# References

http://www.hse.gov.uk/risk/theory/r2p2.pdf

Health and Safety at Work Act of 1974. SI 1974/1439. London. Her Majesty's Stationery Office.

# Appendix B: HAZOP Analysis Example

This appendix provide a hypothetical HAZOP example based on HIPAP 8, HAZOP guidelines published in July 2011, by the State of New South Wales Department of Planning in Sydney, Australia. It details a report of a typical study and serves as a guide for report preparation. The names and minor details have been modified. The original study may be accessed at Haz.hipap8 . rev2011 pdf file.

## Title Page

Title: Hazard and Operability Study (HAZOP) Report, DOP Refineries Ltd., Proposed Project: Distillation Unit at Refinery
Address: 15668 Irene Street, Town, State, U.S. 49614
Chaired by: S. Robinson
Authorized by: J. Mitchell, General Manager of DOP Refineries Ltd.
Authorization date: January 12, 2014

## Contents

Glossary and abbreviations
Summary
HAZOP study
Description of facility
HAZOP team members
HAZOP methodology
Guidewords
Plant overview
Analysis of main findings
Action arising from HAZOP
Minutes
Figure B.1: P&ID
Figure B.2 Revised P&ID

## Glossary and Abbreviations

The application of a formal systematic critical examination of the process and engineering intentions of new or existing facilities to assess the hazard potentials of risks, failures, or malfunction of individual items of equipment and their impacts on the total facility depends on key words:

- Deviation: Departure from design and operating intention.
- Guideword: Tool for visualizing all possible deviations for all design and operating intentions. Broadly speaking, seven kinds of deviation may be associated with a distinctive word or phrase. Guidewords are so named because they guide and stimulate creative thinking about resolving design and operating deviations.
- Techniques and components:
  - Hazard: Deviation that may cause damage, injury, or other form of loss
  - Study team: Small group (normally five to nine people) that performs the study
  - EIS: Environmental impact statement
  - FHA: Final hazard analysis
  - FMEA: Failure modes and effects analysis
  - P&ID: Process and instrumentation diagram
  - PHA: preliminary hazard analysis

## Summary

DOP Refineries Ltd. proposes to construct a refinery for the recovery of kerosene from the waste kerosene solvent returned from auto engine repairers. An environmental impact statement (EIS) and preliminary hazard analysis (PHA) were submitted prior to the approval of the development application (DA). The consent conditions for the DA required that the following study reports be submitted for approval:

- Construction safety
- Fire safety
- HAZOP
- Final hazard analysis
- Transport
- Emergency plan
- Safety management system

The first two studies have been completed and submitted for approval. This report is the third. XYZ Consultants were retained by DOP to provide an independent HAZOP chairman and assist in the preparation of this HAZOP study report.

The prime objective of this HAZOP study was to systematically examine the proposed design and identify, construction issues, hazards, and/or potential operational problems that may be avoided by (mostly minor) redesign or suitable operating procedures before the design is hardened and allows no changes. Selected lines and items in the P&ID were examined by applying appropriate guidewords. The credible unfavorable and potentially hazardous situations and subsequent consequences were evaluated and/or estimated. Measures to eliminate or minimize the undesirable consequences are recommended. The results of the step-by-step procedure and the recommendations were entered in the HAZOP log book. The main HAZOP recommendations are:

> **Recommendations 1 and 2:** (Recommendation numbers used in the minutes have been retained for ease and cross-reference). Install high-flow alarm on raw product feed line to column and a high-level alarm on the column to ensure that operating efficiency is maintained by avoiding the flooding of the reboiler outlet line.
>
> **Recommendations 4, 5, and 12:** Install high-pressure and -temperature alarms on column, furnace, etc., to close the natural gas supply.
>
> **Recommendation 6:** Investigate the need for protection against air suck back into the column on cooling.
>
> **Recommendations 7 and 8:** Investigate the condenser cooling system—water flow, high temperature, high pressure, etc.
>
> **Recommendations 11 and 13:** (a) Consider installing a surge tank in hot oil system to accommodate volume changes due to temperature changes; (b) Investigate possible problems with dead legs and moisture contamination (steam explosions) through venting.

## HAZOP Study

In Chapter 5 discussing HAZOP analysis in detail, we identified a typical format for a report such as this. However, for expediency, fine details are not included in this example. A technical description of the plant, guidewords, and other necessary details are explained briefly to enable readers to follow the meeting minutes and/or log sheets.

## Facility Description

Figure B.1 is a P&ID (Drawing DOP 001, Rev. 1) of the plant. The main items include a distillation column H3, gas fired hot oil furnace H1, product reboiler H2, condenser C1, and associated pumps, controls, and piping. The contaminated waste kerosene is fed to H3 under gravity from a holding tank (not shown). In-flow is controlled by flow control valve VO preset at the desired flow rate. The closed hot oil system uses a heating fluid heated in H1 and circulated through H2 by pump P1. The waste kerosene is boiled in H2 (shell and tube heat exchanger).

A temperature indicator and controller TIC on H3 control the piped natural gas feed valve V1 to the burner in H1 to maintain a set temperature in

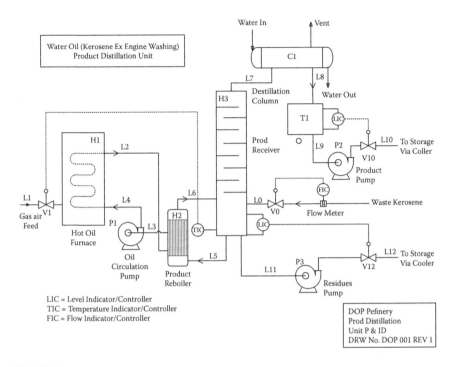

**FIGURE B.1**
Drawing DOP 001, Rev. 1.

H3. The residues in H3 are maintained at the required level by pump P3 and valve V12, which is controlled by the level indicator and controller LIC.

The kerosene vapors in H3 are condensed in C1, a water-cooled shell and tube heat exchanger. A vent is provided to release non-condensables. Level in refined product receiver T1 is maintained by LIC and V10. Product pump P2 transfers product to holding tank (not shown) for distribution to customers by tanker.

## Core HAZOP Team

The HAZOP team was chosen to represent all relevant areas of expertise from design, through commissioning, to operation. The team consisted of the following:

C. Stamatis, design engineer (CS)

J. Robinson, operations engineer (secretary)

K. Huff, maintenance supervisor

S. Robinson, HAZOP chairman

K. Wong, part-time instrument engineer (KW)

## HAZOP Methodology

Selected lines and plant items in the P&ID were examined in turn, starting from L5. [All lines and items are not covered here to conserve space.] Items were recorded in the minutes generally by exception; only key issues likely to pose significant consequences were recorded; see Table B.1. However, items 2 and 3 on minute sheet 1 were included for the purpose of illustration.

Guidewords such as HIGH FLOW (listed below and in minute pages __–__) were applied from a set of guideword cards maintained in a ring binder. Likely causes applicable to each guideword were entered in the second column and credible consequences in the third. The fourth column was for recording existing design or operational safeguards. [No safeguards were found in this simple case.] Where the consequences were likely to present potential hazards or losses involving financial and time impacts, possible changes to the system to eliminate or minimize the consequences were considered and recommendations made. [For simple cases, the recommendation number (Rec #) was noted in the fifth column and the change explained briefly in the sixth column. If several options were presented or further evaluation was required, the recommendations were recorded accordingly.]

## Guidewords Used in Example

FLOW: HIGH, LOW, ZERO, REVERSE

LEVEL: HIGH, LOW

PRESSURE: HIGH

TEMPERATURE: HIGH

CONTAMINATION: e.g., of hot oil heating fluid in contact with moisture in air in surge tank; see Recommendation 11

## Plant Overview

This example covers only the operating mode. In a full HAZOP, where start-up and shutdown procedures are analyzed, more changes may be recommended. The issues to be evaluated further before design changes are made include:

**Recommendation 6:** Consider nitrogen gas purging of H3 and the condenser before start-up to expel air sucked in during cooling for shutdown.

**Recommendation 7:** Recommendations to be adopted after further investigation logged (reported in minutes).

**TABLE B.1**

HAZOP Log Report

| Project: Product Distillation Unit—Waste Oil. Kerosene Exchange Washing | | | | | Node: C1 | | | Page 1 of 2 | |
|---|---|---|---|---|---|---|---|---|---|
| Description: Condenser, water cooled | | | | | | | | Date: September 28, 2012 | |
| | | | | | | | | Drawing No. DOP 001, Rev. 2 | |
| Guideword | Cause | Consequence | Safeguard | Rec # | Recommendation | | | Indiv | Action |
| 1. High Flow | Flow controller fault | Level in column rises, then temperature falls; product reboiler will attempt to maintain temperature in column until reboiler capacity is reached, after which liquid level will rise arid—Mood line LS column stops operating | | 1 | Independent high-flow alarm on L5 | | | KW | |
| 2. Low Flow | 1. Product feed pump failure 2. Jammed isolating valve | Temperature rise and drop in liquid level in column; over-heating; reboiler can handle this; TIC will control gas and air feed to furnace H1; not a problem | | | | | | KW | |
| 3. Zero Flow | As above | As above | | | | | | KW | |

| | | | | | | |
|---|---|---|---|---|---|---|
| 4. High Level | Level controller fault | Flooding of L6 and reboiler stops operation | | 2 | High-level alarm independent, of level controller LIC; alarm level below L6 | KW |
| 5. Low Level | Level controller malfunction or low flow | Not a problem (as for no flow) | | 3 | Low-level alarm | KW |
| 6. High Pressure | Water failure in condenser | Condenser vent acts as relief device; no adverse effect | | 4 | Pressure indicator on column; high-pressure alarm and trip on gas/air control valve V1 | KW |
| 7. High Temperature | Loss of feed | No adverse effect | | 5 | High- and low-temperature alarm on TIC; additional high-temperature alarm linked to furnace gas inlet shutoff | KW |
| 8. Reverse Flow | Cooling of condenser and H3 after shutdown | Suck-back of air into H3 on cooling | | 6 | Consider nitrogen purge | CS |
| 9. High Pressure | Water failed or flow | Excess pressure | | 7 | Backup cooling water system; thermocouple on vent; reorientation of water line for countercurrent flow | CS |
| 10. High Level | Pump P2, LIC, or V10 fault | T1 overfills; high pressure in C1 and H3 if C1 floods | | 8 | LAH (independent) on T1 | KW |
| 11. Low Level | LIC or V10 fault | Pump damage | | 9 | Consider LAL (independent) | KW/CS |

*Continued*

**TABLE B.1 (Continued)**

HAZOP Log Report

| Guideword | Cause | Consequence | Safeguard | Rec # | Recommendation | Indiv | Action |
|---|---|---|---|---|---|---|---|
| 12. Low Flow | P1 fails | Loss of heat to H2; TIC will call for further opening of V1, increasing temperature in H1 | | 10 | Install flow sensor/indicator or alarm to trip furnace via V1 or other | KW | |
| 13. High Pressure | Heating; expansion of hot oil | Burst pipes, etc. | | 11 | Surge tank in oil system; evaluate location of tank: on L3 (at pump suction) or L2; check dead leg and moisture condensation in oil | CS | |
| 14. High Temperature | 1. High product load on H3 causing high flame in H1<br>2. TIC on H3 failed, V1 failed to open<br>3. H2 partly blocked or heat transfer poor | High temperature in furnace | | 12 | Pyrometer in furnace to alarm or trip gas supply | CS | |
| 15. Contamination (water in oil) | Water from atmosphere through vent | Water turns to steam and explodes | | 13 | Locate surge tank in hot system; avoid dead legs; steam vents at high points in pipe system; nitrogen connection on vent | CS | |

**Recommendation 9:** Consider installing LEVEL LOW alarm on product receiver to trip P2 against damage from running dry.

**Recommendations 11 and 13:** Consider installing surge tank on hot oil system to accommodate expansion. Location to be decided after considering effects of dead legs, moisture, etc. Consider nitrogen padding to eliminate condensation.

## Analysis of Main Findings

The main findings were evaluated by adopting the following methodology:

1. The outcome of each deviation was evaluated to verify whether the consequence would pose a hazardous condition to the plant or those within and outside the site.
2. Conditions likely to cause frequent loss of production were also included.
3. If a hazardous or loss scenario appeared credible, the analysis was continued to develop a safeguard to eliminate or minimize the possibility.
4. Where the possibility still existed (although reduced), additional alarms and trip systems were recommended.

The study results are detailed in the minutes on pages __–__. The recommendations arising from the study are:

**Recommendation 1:** Install high-flow alarm on L0. A flow controller fault may signal valve V0 to pass more than the necessary quantity, resulting in flooding of L6 and slowing the heating process. Although adverse effects are unlikely, the poor operation could be improved by installing a high-flow alarm for early operator intervention.

**Recommendation 2:** Install high-level alarm (independent of level controller LIC) in column H3. A level controller fault may cause flooding of L6 and slowing of operation (see above). An independent alarm at a level above the normal control level and below L6 could alert an operator to take early action.

**Recommendation 3:** Install low-level alarm (not necessarily independent of LIC) on column H3 to ensure early operator intervention and avoid production losses. No adverse consequences.

**Recommendation 4:** Install pressure indicator and high-pressure alarm on column H3 to close the natural gas supply valve V1 to furnace H1. High pressure in column H3 may arise from several causes, one of which is the failure of cooling water supply to the condenser. Although the condenser vent will act as a relief valve, this is not desirable.

**Recommendation 5:** Install high- and low-temperature alarm on TIC on column H3 to alert operator of malfunction. Additional independent high-temperature alarm to be installed to shut natural gas supply valve V1 to furnace H1. No immediate adverse effects are likely with temperature rise. However, it was considered prudent to shut the gas supply to avoid unnecessary overheating of reboiler H2 tubes if kerosene level in reboiler falls too low.

**Recommendation 6:** Further investigation is recommended into the possibility of air suck-back through the condenser vent when column H3 cools after shutdown. The air can cause corrosion in column H3 and also form explosive mixtures with the kerosene vapor on start-up. A nitrogen purge system should be considered.

**Recommendation 7:** Investigate need for a back-up cooling water system for condenser C1, a thermocouple on condenser vent, and reorienting condenser water lines for countercurrent and bottom-in/top-out flow. A loss of cooling water will produce high pressures in the condenser and column H3. The thermocouple at the vent will provide early warning of low water flow rate.

**Recommendation 8:** Install high-level alarm (independent of LIC) on product receiver T1 to avoid overfilling and subsequent over-pressure in condenser and column H3 from failure in product pumping system (P2, V1, LIC, etc.). Investigate whether alarm should be audible for operator intervention or automatically shut down gas supply to furnace H1.

**Recommendation 9:** Investigate need for low-level alarm or pump trip in the event of level controller fault in T1 to avoid pump damage from running dry.

**Recommendation 10:** Install flow sensor, indicator, or alarm on hot oil circulation lines to shut gas supply to furnace H1 and avoid temperature rise if loss of flow occurs in circulation system.

**Recommendation 11:** Consider installing a surge tank in hot oil system to accommodate volume changes due to temperature changes. Evaluate location of surge tank at pump suction or on L2. Check effects of dead leg and moisture (condensation) in oil.

**Recommendation 12:** Install a pyrometer in furnace to sound alarm and shut gas supply to furnace H1 if high temperature in H1 occurs from loss of hot oil, for example, from pipe leak, TIC fails to close V1, or poor heat transfer occurs in H2.

**Recommendation 13:** The surge tank in Recommendation 11 must be vented. Condensation of moisture on cooling can contaminate the oil and cause steam explosions on reheating. Consider nitrogen padding and steam vents at high points.

Consequence and/or risk analysis was considered necessary for the issues raised by Recommendations 6, 7, 9 11, and 13. Detailed analysis subsequent to the HAZOP indicated that:

> **Recommendation 6:** A continuous bleed of nitrogen should be maintained to the condenser or the column H3 on system shutdown, in order to prevent air suck back on cooling.
>
> **Recommendation 7:** Reorientation of the condenser water lines as recommended and the installation of a thermocouple at the vent were adequate to ensure the necessary level of integrity of the condenser system. A backup cooling water system was not justified.
>
> **Recommendation 9:** A no-flow switch on the pump delivery would protect the pump against dry operation due to LIC or V10 faults and in addition provide protection against no-flow from any other causes such as blocked lines, inadvertently closed valves, etc.
>
> **Recommendations 11 and 13:** The surge tank should be installed on line L2 on the outlet side of the hot oil furnace, together with continuous nitrogen padding to prevent ingress of moisture.

## Actions Arising from HAZOP

The recommendations modified by the outcomes of subsequent detailed analyses cited above have all been incorporated into the design as shown on the revised P&ID DOP 001, Rev. 2 (Figure B.2). Note the changes to the hot oil system, condenser water flow, and additional instruments and alarms. The implementation of the changes as a result of the HAZOP has not revealed any actions that may be considered potentially hazardous to plant personnel, the public, or the environment.

The pre-commissioning and commissioning checklists and test procedures have been modified to ensure that the final HAZOP recommendations are verified at all appropriate stages based on Table B.1.

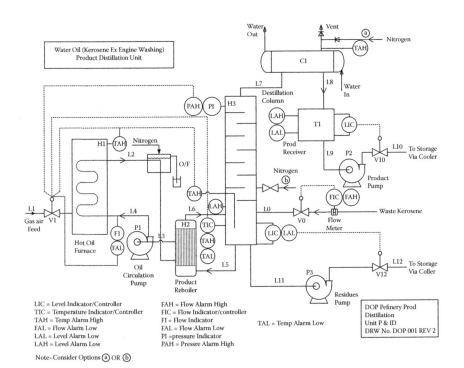

**FIGURE B.2**
Drawing DOP 001, Rev. 2.

## Selected Bibliography

Aksorn, T. and B. H. W. Hadikusumo (2008). Measuring the effectiveness of safety programmes in the Thai construction industry. *Construction Management and Economics*, 26, 409–421.

Andow, P. (1991). *Guidance on HAZOP Procedures for Computer-Controlled Plants*. HSE Contract Research Report 26/1991. London: Her Majesty's Stationary Office.

Balemans, A. (1974). Checklist guide lines for safe design of process plants. *First International Loss Prevention Symposium*. London.

Bancroft, K. (2002). Job hazard analysis for unsafe acts. *Occupational Health and Safety*, 71, 206–215.

Beach, L. (1992). *Image Theory: Decision Making in Personal and Organizational Contexts*. New York: John Wiley & Sons.

BS5760:2 (1992). Guide to the Assessment of Reliability.

BS5760:5 (1992). Guide to Failure Modes, Effects, and Criticality Analysis (FMEA and FMECA).

BS5760:7 (1992). Guide to Fault Tree Analysis.

BS5760:9 (1992). Reliability of Systems, Equipment, and Components. Part 9: Guide to Block Diagram Technique.

Chastain, J. and J. Jenson (1997). Conduct better maintenance and operability studies. *Chemical Engineering Progress*, 93, 49–53.

Clemens, P. and T. Pfitzer (2006). Risk assessment and control. *Professional Safety*, 51, 41–44.

Elmendorf, M. (1996). Introduction to process hazard analysis. *Journal of Environmental Law and Practice*, 4, 36–56.

Fairley, R. (1985). *Software Engineering Concepts*. New York: McGraw-Hill.

Garvey, P. (2008). *Analytical Methods for Risk Management: A Systems Engineering Perspective*. Boca Raton, FL: Taylor & Francis.

Garvey, P. (2000). *Probability Methods for Cost Uncertainty Analysis: A Systems Engineering Perspective*. Boca Raton, FL: Taylor & Francis.

Geronsin, R. (2001). Job hazard assessment: a comprehensive approach. *Professional Safety*, 46, 23–30.

Giebe, K. (2013). Zero injuries: how safety and productivity find common ground. *Fabricating and Metal Working*, Sept., 54–55.

Gould, J., M. Glossop, and A. Ioannides (2000). *Review of Hazard Identification Techniques*. Health and Safety Laboratory. HSL/2005/58.

Greenberg, H. and J. Cramer (1992). *Risk Assessment and Risk Management for the Chemical Process Industry*. New York: Van Nostrand Reinhold.

Haddad, S. (2011). *HIPAP 6: Hazard Analysis*. Sydney: New South Wales Department of Planning.

Hammer, W. (1989). *Occupational Safety Management and Engineering*, 4th ed. Englewood Cliffs, NJ: Prentice Hall.

Hoxie, W. (2003). Preconstruction risk assessments. *Professional Safety*, 48, 50–53.

http://www.hse.gov.uk/risk/theory/r2p2.pdf

http://www.stb07.com/process-safety-management/process-hazards-analysis.html

Kossiakoff, A. and W. N. Sweet (2003). *Systems Engineering Principles and Practice.* New York: John Wiley & Sons, pp. 98–106.

Morris, J. and J. Wachs (2003). Implementing a job hazard analysis program. *AAOHN Journal,* 51, 187–193.

Occupational Safety and Health Administration. (2002). *Job Hazard Analysis* (OSHA 3071).

Olsen, R. (2011). Activity Hazard Analysis in a Marine Construction Environment. Master's thesis. University of Wisconsin. http://www2.uwstout.edu/content/lib/thesis/2011/2011olsenr.pdf.

Palmer, J. P. (2004). Evaluating and assessing process hazard analyses. *Journal of Hazardous Materials,* 115, 181–192.

Rasmussen, B. and C. Whetton (1997). Hazard identification based on plant functional modeling. *Reliability Engineering and Systems Safety.* 55, 77–84.

Rozenfeld, O., R. Sacks, Y. Rosenfeld et al. (2010). Construction job safety analysis. *Safety Science,* 48, 491–498.

Smith, K. and D. Whittle (2001). Six steps to effectively update and revalidate PHAs. *Chemical Engineering Progress,* 97, 70–77.

Stewart, M. and R. Melchers (1997). *Probabilistic Risk Assessment of Engineering Systems.* New York: Chapman & Hall.

Swartz, G. (2002). Job hazard analysis. *Professional Safety,* 47, 27–33.

Topping, M. (2001). Role of occupational exposure limits in the control of workplace exposure to chemicals. *Occupational and Environmental Medicine.* 58, 138–144.

U.S. Army Corps of Engineers (2008). *Safety and Health Requirements Manual,* EM 385-1-1. Washington, DC: Government Printing Office.

U.S. Department of the Army (2006). *Composite Risk Management,* FM 100-14. Washington, DC: Government Printing Office.

U.S. Department of the Army (2010). *Army Safety Program,* Pamphlet 385-10. Washington, DC: Government Printing Office.

Vaughan, E. and M. Seifert (1992). Variability in the framing of risk issues. *Journal of Social Issues,* 48, 119.

Vincoli, J. W. (1993). *Basic Guide to System Safety,* New York: Van Nostrand Reinhold.

Wells, G., M. Wardman, and C. Whetton (1993). Preliminary safety analysis. *Journal of Loss Prevention in the Process Industries,* 6, 47–60.

# Index

## A

Accident history, 179
Accident investigation, 149, 201
Accountability, 27
Accredited engineers, 116
Actions arising from HAZOP, 225
Adjacent facilities, 58
Agenda, 168, 170–171, 173
AHA. *See* Australian Health
    Administration
ALARP. *See* As low as reasonably
    practicable
ALARP fallacies, 37–39
Alternative possible outcomes, 13, 25
Alternative worlds, 13, 25
Analysis of main findings, 223–225
As low as reasonably practicable, 27–39
Atmosphere, 78, 91, 105, 170, 222
Audit risk, 10
Audits, 18, 53, 146, 150, 201
Australian Health Administration, 202
    checklist, 202
    guidelines, 202
Automation engineer, 116

## B

Bell Telephone Laboratories, 125
Block diagram, 134–136
Brainstorming, 72, 78, 83, 86, 90, 92, 99,
    106, 144, 174
Buried cables, 58
Business interruption, 40, 144
Business risks, 4

## C

CCTA risk analysis and management
    method, 8, 17
Certified and accredited engineers, 116
CFR 910.1200
    communication, 187–200
    employee access, 190

employee training, 189–190
information sources, 191–200
    chemical materials lists, 193
    labels, 191
    material safety data sheets,
        193–200
    safe use instruction, 191–193
    members, 188–189
    responsibilities, 189
Chairperson, 96, 98–100, 102, 106, 111,
    113, 116, 165, 174
CHAZOP. *See* Computer HAZOP
Chebyshev inequality law, 14
Chebyshev's inequality law, 14
Checklist, 61–64
    chemical storage, 63–64
    questions, 62
Checklists, 201–213
Chemical storage checklist, 63–64
Civil engineer, 4, 116
Cold War, 2
Commitment, 166, 175, 177, 179, 201
Component failure, 87
Computer HAZOP, 108–110
Consensus, 165–167
Construction material toxicity, 58
Contents, HAZOP analysis, 215
Control of hazardous energy, 31
Control plan, 137–138
Core HAZOP team, HAZOP analysis,
    218
Core team, 116
Corporate image impact, 144
Cost-benefit analysis, 34
CPT. *See* Cumulative prospect theory
CRAMM. *See* CCTA risk analysis and
    management method
Cranes, 58, 61
CRIOP. *See* Crisis intervention in
    offshore production
Crisis intervention in offshore
    production, 146
Cultural alignment, 27

Cultural change, 27
Cumulative incidence, 3, 15–16
    measure of frequency, 15
Cumulative prospect theory, 19
Cyber warfare, 4

**D**

Defining risk, 1
Definition of risk, 1, 86, 160
The Delta Works, 19
Demolition safety, 58
Department of Labor, 177, 200
Deriving recommendations, 106
Design engineer, 100, 218
DETAM. *See* Dynamic event tree
    analysis method
Differentiating risks, 40–43
Difficult team members, 167–168
Digraph, 54–55
Documented safety philosophy, 201
Dow Chemical Exposure Index, 65
Dow Fire and Explosion Index, 65
Drains, 58, 75, 94, 120, 210
DYLAM. *See* Dynamic event logic
    analytical methodology
Dynamic event logic analytical
    methodology, 56
Dynamic event tree analysis method,
    56–57
Dynamic systems, methodologies for
    analysis, 54–57

**E**

Economic risks, 3
Effluents, 58, 113
Electrical classification areas, 58
Electrical engineer, 100–101, 116
Electrical systems design, 31
Electrocutions, 31
Emergency environment event, 88
Emergency shutdown systems, 58
Employee access, 190
Employee training, 189–190
Engineers, 100–101
Environmental impacts, 144
Environmental Protection Agency, 2
ETA. *See* Event tree analysis

Event tree analysis, 150–159
Example of hazard and operability
        analysis, 215–228
    abbreviations, 215–216
    accident investigation, 149
    audits, 150
    block diagram, 134–136
    contents, 215
    control plan, 137–138
    core HAZOP team, 218
    crisis intervention in offshore
        production, 146
    event tree analysis, 150–159
    facility description, 217–218
    failure mode analysis, 137
    failure mode and critical analysis,
        141
    functional flow diagram, 134–136
    glossary, 215–216
    hazard analysis and critical control
        points, 146–148
    hazard identification, 142–146
    human reliability analysis, 140–141
    layouts, and schematics, 136–137
    methodologies, 133–160
    near-miss reporting, 148
    need and feasibility analysis, 138
    process flowchart, 133–134
    process potential study, 138
    recommendations, 217
    schematics, 136–137
    semi-quantitative risk assessment,
        150
    sketches, 136–137
    summary, 216–217
    task analysis, 138–140
    title page, 215
*Exxon Valdez*, 2

**F**

Facility description, 217
    HAZOP analysis, 217–218
Facility location checklist, 202, 205–213
Failure mode analysis, 137
Failure Mode and Critically Effects
        Analysis, 46–51
Fall protection, 31
Falls, 31, 73, 220, 224

Fatality, 31, 33–34, 39–40, 71, 80, 82
  financial cost of, 34
Fault graph, 54–55
Fault tree analysis, 125–132
  Bell Telephone Laboratories, 125
  benefits, 128
  construction rules, 128–132
  reliability block diagrams, 125
Fear as intuitive risk assessment, 9–10
Feedback, 27, 55, 73, 139, 148, 173
Flares, 58, 60, 206, 208–209
FMA. *See* Failure mode analysis
FMCEA. *See* Failure Mode and Critically
    Effects Analysis
FMEA documentation, 47
FMEA generation flow, 48
Focus on incident control, 27
Fog, 58, 195, 208–209
Follow-up, 108
Forklifts, 58
Framing, 7, 16, 21–22
Freezing, 58, 73
FTA. *See* Fault tree analysis
Functional flow diagram, 134–136

**G**

Generating FMEA, flow, 48
Glossary, HAZOP analysis, 215–216
GO method, 54
Groupthink, 7
Guidelines, meeting management,
    174–175
Guidewords, 46, 60, 83–85, 91–93, 96–99,
    102–104, 110–112, 114, 117,
    215–217, 219

**H**

Hazard and operability analysis, 86–90
  defining risk, 86–87
  definition, 90–91
  description of process, 101–110
  detailed analysis, 95–99
  deviations from design intent, 97–98
  documentation and follow-up, 93–95
  effectiveness factors, 99
  examination, 92–93
  minimum requirements, 86

preparation, 91–92
process, 86–95
report, 110–114
review, 114–119
sequence of examination, 96–97
success factors, 119–121
team, 99–101
trigger events, 87–88
use of analysis, 88–90
Hazard and operability studies, 88, 98,
    110–111, 118–119, 144, 215–217
Hazard communication program,
    188–200
Hazard communication standard, 31,
    187, 193
Hazard identification, 142–146
Hazardous materials control committee,
    188–200
Hazardous materials control system,
    191, 193
HAZID. *See* Hazard identification
HAZOP/FMEA worksheet, 49
HAZOP methodology, 83–84, 107, 215,
    219
HAZOP team, 164–165
Health and Safety at Work Act, 28, 213
High reliability organizations, 5–6
High temperatures, 126
HMCS. *See* Hazardous materials control
    system
Hoists, 58
HROs. *See* High reliability
    organizations
Human factors risks, 7–8
Human failure, 87, 141
Human reliability analysis, 140–141
Human services risks, 5

**I**

Ice, 58, 76, 172
Illness, 3, 34, 71–72, 80, 85, 177–179, 183
  financial cost of, 34
In-house safety committee, 201
In-process meeting management,
    172–173
Incidence proportion, 3, 15–16
Incident control focus, 27
Indexing, 65

Industrial trucks, 31
Inequality law, Chebyshev, 14
Information assurance, 4
Information security, 2, 4, 18, 21
Information sources, 191–200
    chemical materials lists, 193
    labels, 191
    material safety data sheets, 193–200
    safe use instruction, 191–193
Information technology risks, 4
Input documents, 114–115
Instrument disturbance, 88
Insurance risks, 4
Interface hazards analysis, 65–67
International Organization for
        Standardization, 15, 17, 21
Investigation, 40, 50, 53, 117, 149, 164–165,
        201, 219, 224
ISO. *See* International Organization for
        Standardization
IT risk, 4

**J**

Job review, 179

**L**

Ladders, 31
Launching team, 116
Layouts, 136–137
Lead process engineer, 116
Leader, 116
Leaks, 58, 60, 74, 192
Lifetime risk, 3, 16
Line responsibility, 201
Lock-out, 31
Low as reasonably practicable, 5, 25, 27

**M**

Machine guarding, 31
    machines, 31
Magement risks, 4
Maintenance procedures, 58
Major risks, 40–41
Management commitment, 27, 201

Marine readiness, 85
Markov analysis, 55–56
Mechanical engineer, 116
Meeting planning, 170–172
    agenda, 170
    atmosphere, 170
    costs, 170
    follow-up, 171
    participants, 170
    post-meeting work, 170
    purpose, 170
    time considerations, 170
Meeting questions, 107–108
Meeting records, 106–107
MEHARI. *See* Method for harmonized
        analysis of risk
Members, 188–189
Method for harmonized analysis of
        risk, 8
Methodologies, 46–67
Mid-level managers, 164–165
Middle Ages, 1
Middle Eastern traders, 1
Minor risks, 41
Motivation, 201

**N**

National Institute of Standards and
        Technology, 17
NIST. *See* National Institute of
        Standards and Technology
Noise, 30, 58, 61, 72, 104, 113, 144,
        171–172, 178, 206, 208, 212
North African Arab traders, 1
Nuclear Regulatory Commission, 6, 21,
        160

**O**

Occupational Health and Safety
        Advisory Services, 2
Occupational Safety and Health
        Administration, 177–186, 189
    accident history, 179
    employee involvement, 179
    preliminary job review, 179

The Open Group, 15
Operating procedures, 58, 60, 95, 101, 162, 165, 180, 203, 216
Operations manager, 100–101
OSHA, 30–31, 58, 65, 86, 177–179, 181, 184–186, 189, 195, 200
Overhead cables, 58

**P**

Part-time team, 116
Participants, 113, 115–117, 170, 172–175, 203
Perimeter fencing, 58
Permanently incapacitating injury, 34
PHA. *See* Preliminary hazard analysis
Philosophy, 7, 26, 37, 85, 161, 201–202
Pipeline Risk Management Index, 65
Piping engineer, 116
Piping systems, 58, 60, 188
Pitfalls, meeting, 173–174
Planning meetings, 170–172
    agenda, 170
    atmosphere, 170
    costs, 170
    follow-up, 171
    participants, 170
    post-meeting work, 170
    purpose, 170
    time considerations, 170
Plant overview, 215, 219
Point of reference concept, 102–104
Powered industrial trucks, 31
PPS. *See* Process potential study
PRA. *See* Probabilistic risk assessment
Pre-meeting work, 170
Preliminary hazard analysis, 69–82
    corrective measures, 81–82
    limitations, 81
    preventive measures, 81–82
    probability, 80–81
    severity, 80–81
Preliminary job review, 179
Probabilistic risk assessment, 6, 54, 160
Problem solving, 169–170
Process engineer, 100, 116
Process flowchart, 133–134

Process potential study, 138
Project engineer, 115–117
Property damage, 40, 71, 80, 144
Prospect theory, 12, 19, 22
Public access, 58

**Q**

Qualitative methodologies, 45–51
    failure mode and effects analysis, 46–51
    hazard and operability studies, 46
    preliminary risk analysis, 45–46
Quality impacts, 144
Quantitative analysis, 8–9

**R**

Rain, 58, 172
Rate of ruin, 12, 19
RBDs. *See* Reliability block diagrams
*The Reactor Safety Study*, 6
Recommendations, HAZOP analysis, 217
Reliability block diagrams, 135
Replaced by State-of-the-Art Reactor Consequence Analyses, 6
Report, hazard and operability analysis, 110–114
    abbreviations, 111
    aim, 111
    description of facility, 111–113
    findings, 114
    glossary, 111
    guidewords, 111
    main findings analysis, 114
    methodology, 113
    overview, 113–114
    recommendations, 111
    scope of report, 111
    study title page, 110–111
    summary of main findings, 111
    table of contents, 111
    team members, 113
Respiratory protection, 31, 195
Responsibilities, 189
Review information pack, 115

Review team composition, 115–117
Risk, uncertainty, distinguished, 10–11
Risk assessment analysis, 8
Risk assessment matrix, 43
  categories of, 43
Risk attitude, 11–12
Risk aversion, 7, 16
Risk leverage, 36–37
Risk priorities, 41–43
Rotary engineer, 116
Rules and procedures, 201

**S**

Safety communications, 201
Safety goals, 201
Safety philosophy, 201
Safety plan checklist, 201–204
Safety staff, 201
Safety training, 201, 203
Scaffolding, 31
Scenario analysis, 2, 13–15, 25
Secretary, 116
Security risks, 6
Semi-quantitative risk assessment, 150
Senior managers, 165
Serious injury, financial cost, 34
Serious risk, 41
Shared vision, 27
Shutdown, 33, 43, 58, 76, 89, 103–104, 109,
  112–113, 143, 219, 221, 224–225
Sketches, 136–137
  layouts, 136–137
Snow, 76, 172
Societal risks, 6–7
Soviet Union, United States,
  confrontations, 2
SQRA. *See* Semi-quantitative risk
  assessment
Standard CFR 910.1200
  communication, 187–200
  employee access, 190
  employee training, 189–190
  hazard communication program,
    188–200
  hazardous materials control
    committee, 188–200

  information sources, 191–200
    chemical materials lists, 193
    labels, 191
    material safety data sheets, 193–200
    safe use instruction, 191–193
  members, 188–189
  responsibilities, 189
State-of-the-Art Reactor Consequence
    Analyses, 20
Summary, HAZOP analysis, 216–217
Supply failure, 87
Supportive safety staff, 201

**T**

Tag-out, 31
Task analysis, 138–140
Team leader, 100
Team overview, 162
Team process check, 167–168
Teams, 161–175
  benefits, 163–164
  consensus, 165–167
  difficult team members, 167–168
  HAZOP team, 164–165
  in-process meeting management,
    172–173
  meeting management guidelines,
    174–175
  meeting planning, 170–172
    agenda, 170
    atmosphere, 170
    costs, 170
    follow-up, 171
    participants, 170
    post-meeting work, 170
    pre-meeting work, 170
    purpose, 170
    time considerations, 170
  mid-level managers, 164–165
  pitfalls, 173–174
  problem solving, 169–170
  senior managers, 165
  team process check, 167–168
  technicians, 164
Technicians, 164
Temperatures, 126, 178, 207

"Theory of Leaky Modules," 7
Tides, 58
Title page, HAZOP analysis, 215
Toxicity, 58, 104
Traditional methodologies, 57–67
Tree-based techniques, 52–54
    cause-consequence analysis, 53
    event tree analysis, 52
    fault tree analysis, 52
    management oversight risk tree
        analysis, 53
    safety management organization
        review technique, 53
Type of injury, explanation, value, 34

**U**

Uncertainty, risk, distinguished, 10–11
United States, Soviet Union,
        confrontations, 2
Upstream systems, 27
Upstream systems definitions, 27
U.S. Department of Labor's
        Occupational Safety and
        Health Administration, 177

U.S. Environmental Protection Agency, 2
U.S. Statistics, 30–31

**V**

Vector quantity, risk as, 12
Vents, 58, 60, 94, 120, 222, 224

**W**

Weather problems, 58
What-if analysis, 58–61
Winterization, 58
Wiring, 31, 126
    methods, 31

**Y**

Young people, work accidents involving,
    30

**Z**

Zero mind-set, 25–27, 85